6/2

For David Gauvin (french pronunciation),
Scientist extraordinaire,
Lover of near-crustacean
mentalities, and
Dear partner in Gossip.

With love,

From your unemployed
scientist friend,

Lauren

Pigeon Lore

Pigeon Lore

A. NEILSON HUTTON

FABER AND FABER LIMITED
3 Queen Square, London

This new edition first published in 1978
by Faber and Faber Limited
3 Queen Square London WC1
Printed in Great Britain by
Latimer Trend & Company Ltd Plymouth

Publisher's Note

Owing to illness, the author was unable to correct the final proofs of this new edition, publication of which has been considerably delayed. The publishers are grateful to Colin Osman for help in seeing the book through the press.

British Library Cataloguing in Publication Data

Hutton, Andrew Neilson
Pigeon Lore. – New ed.
1. Pigeon racing
I. Title
798'.8 SF469

ISBN 0–571–10409–6

Contents

Introduction

The evident popularity of *Pigeon Racing: Win With Olympic*, the Danish edition, *Brevduesporten Krav*, and the urgings of many friends for me to put my experiences into print encouraged me to write this book.

I do not remember being without pigeons and have witnessed many changes in the sport of pigeon racing, but despite the progress made it is remarkable how old beliefs and prejudices linger on. In my early days there was a strong prejudice against those who kept pigeons, traces of which remain to this day and find expression in many odd ways.

Reconsidering all the circumstances I now realize that among the many things for which I have reason to be grateful to my parents, was their tactful handling of my passion for pigeons. Their control over my pigeon activities demanded my observance of two simple rules; the first that I should not keep company with anyone I could not bring to the house to meet the family; the second, any pigeons I wanted had to be paid for from my own savings. I recall that I paid for my first expensive purchases from the profit made from finding and supplying broody hens to a poultry-keeper.

The most serious objection my parents had was to anyone calling to see my pigeons on a Sunday, still a popular day for visiting by fanciers. At that time, Sunday liberations were taboo; and although, later on, the taboo was lifted a little, clocks could not be checked until the Monday.

Within the pages of this book much is said about the effects of environment, and I have not the slightest doubt that each of us is influenced by our early associates and experiences. My earliest associate was my near neighbour and lifelong friend, the late Bob Hamilton of Whifflet, Coatbridge, in the days when unrung pigeons identified by wing stamps could be raced and we timed in by telegram. Those early clubs had wide radii and Bob and I tramped miles, visiting fellow members. When eventually a local club was formed, with Bob as secretary, I remember how he and I wrestled with the intricacies of clock variations and velocities. Another early associate was the late Matt Greenshields of Salsburgh, who introduced me to the niceties of pigeon lore and the importance of observing the little details, thus whetting my appetite to know more and to finding the reason why.

With the passing of years came the ups and downs of pigeon racing; the beginning of a long and close friendship with the late Jack T. Clark of Windermere; and the friendly rivalry in racing and debate with the members of the Lanarkshire Social Circle—a hard school to which many came but few stayed. The late Dr. William Anderson declared it to be the hardest club in Scotland in which to win.

On the administrative side I have served in many capacities, including a term as President of the Scottish Homing Union, and as a member of the National Pigeon Service Committee, an Advisory Committee to the Minister of State for Air. About the teething troubles of those hectic and exciting days during the formation of the National Pigeon Service (civilian), supply of pigeons to the Royal Air Force, I could fill a book. My most lasting impression of them is the sterling work of J. Selby-Thomas as secretary to the committee.

Later, I had the good fortune to be appointed commander of the Pigeon Service Holding Unit, through which the majority of officers and men of the Army Pigeon Service passed for initial training, and later for refresher courses prior to being mobilized and equipped for service overseas.

Here I had the opportunity of experimenting with my theory that led to the development of fresh techniques in the use of pigeons as carriers of messages. Carried out in a variety of circumstances not encountered under normal conditions of training and racing, these experiments revealed many hitherto unsuspected aspects in the behaviour of pigeons. Much of the credit for their success is due to the hard work and patience of Sergeant Alex B. King of Wishaw. The results obtained set me thinking furiously again in an endeavour to adjust my ideas to fresh concepts of racing pigeons and what motivates their actions in given circumstances. These fresh concepts of pigeon behaviour were put to practical use in the management of the 'Boomerang System' of two-way message carrying and mobile loft technique. Trained in these, the late Harry Smith, BEM, carried on the good work in the Middle East.

At their own request the RAF sent an NCO to the Army Pigeon Service Holding Unit for a course of training in the new techniques.

With the 1st Army I had the satisfaction of putting these new techniques into practice under active service conditions in Algeria and Tunisia. During these operations I was loyally supported by Sergeants Arthur Stockdale and James B. Agnew, and the ever-faithful Charlie Stafford. Captain Roland Smith and his merry men of 14 HQ Section (CP) were a tower of strength under difficult conditions. In the short period of six months pigeons successfully carried messages to seven different loft locations. The first news of the capture of Tunis was carried by pigeon.

Based on my own experience I have set down my observations in the hope that they may prove helpful to those who follow. It is said that 'you cannot teach an old dog new tricks' and I know very well how hard it is to discard the thoughts of a lifetime for something new. That being so, the old hand may find it difficult to think again to the point of accepting and practising new ideas. Beginners, on the other hand, by reason of their enthusiasm, tend to be over-credulous in accepting what they read and hear as facts, rather than often only opinion.

I wish to thank those who have so willingly supplied me with material and checked details I have used in the text but who are not specifically mentioned therein.

I am especially indebted to Wendell M. Levi, Sumter, South Carolina, USA, author of that masterpiece *The Pigeon*, for his most helpful suggestions and encouragement during the period

I was preparing the original manuscript for the first edition; and also to the novices whose awkward questions often caused me to think again.

A. NEILSON HUTTON

Glasgow
October 1961

INTRODUCTION TO SECOND EDITION

For the second edition I decided to take the opportunity of revising parts of the text and adding fresh information of interest together with up-to-date illustrations. The letters I receive are usually enlightening and have helped me in preparing this second edition. I do not mind constructive criticism, having found in retrospect that it has been from those who have told me I was wrong that I obtained the most benefit from having to 'think again'. I would like to take this opportunity of stating that letters which I receive are treated as confidential and will not be used for publicity purposes without the writer's permission. I am gratified to know that so many have bought my books through being recommended to do so by another fancier. Chapter 9, 'Inheritance', has created a great deal of interest. It is, however, unfortunate that the differences in the nomenclature used by fanciers in describing their pigeons lead to misunderstandings and pointless controversy. It is also evident that some writers have not made full use of the index and glossary. To help the reader I have included at the back of this book Standard Nomenclature—a brief description of how to differentiate between the most common colours and plumage markings.

A. NEILSON HUTTON

Glasgow
July 1966

INTRODUCTION TO THIRD EDITION

Letters received from readers of the first two editions of *Pigeon Lore* led me to believe that it is a book that serves a very useful purpose; and when it became evident that if the book was to stay in print a new edition was necessary, I undertook to revise it and bring it up to date. While preparing this third edition, I kept very much in mind the younger generation, since they look to the future. Thus, Chapter 6 has been revised by the deletion of material that would not particularly interest the young pigeon fancier; and the chapter on 'Pioneers' has been deleted and replaced by a new Chapter 8, 'Second Thoughts', because I felt that this belonged to a generation of pigeon fanciers that, sadly, is becoming smaller. Chapter 9, too, has been completely rewritten in the light of fresh discoveries about the genetic constitution of birds, especially pigeons, as compared with mammals. Appendix B has been replaced and deals now only with the essential facts of sanitation in the pigeon loft; and I have added a fourth appendix—Appendix D—which discusses the problems of strays which, like the poor, are always with us.

It is said that advice, especially good advice, is a waste of time; but I will risk this criticism by warning readers that before they become involved in controversy they should make sure that the source of their information is up to date. I can't stress too much the importance of making the fullest use of the Standard Nomenclature, Glossary and Index, all of which will be found at the end of the book.

A. NEILSON HUTTON

Glasgow
1973

1 Stimuli to Behaviour

The use of pigeons to carry messages can be traced back into the mists of antiquity; the type of pigeons used seems to have varied with the period. Whatever the type may have been, one thing is certain, they all possessed the ability to home to their lofts. The formation of pigeon racing clubs in Belgium at the beginning of the nineteenth century gave an impetus to improving the performance of the Cumulet and Smerle, and possibly others of a similar type, in Belgium; and in Great Britain the Dragoon. The cross-breeding of these and the severe selection of over one hundred years of competitive racing have moulded the racing pigeon as we know it today. A contemporary writer claims that the modern racing pigeon, far from being the product of cross-mating, is in fact the direct product of the Rock Dove itself. In support of his contention, he draws attention to the remarkable similarity between the Rock Dove and the racing pigeon in appearance, measurements, weight and habits. I agree that the racing pigeon more closely resembles the Rock Dove than did its progenitors the Smerle, Cumulet and Dragoon. Furthermore, the evidence we have points to these in their turn being the product of other varieties, now extinct, or so cultivated by man for exhibition purposes that they do not now resemble the originals. The reason for this is that if we cross-breed at random several varieties of pigeons, the resultant progeny quickly lose their distinctive physical variety characteristics, and in three or four generations come to resemble a common ancestor type. Therefore, the survival of the distinctive physical characters of the fancy varieties is entirely dependent on the care in selection for mating.

The racing pigeon was primarily bred for performance in racing and, irrespective of what the old pioneers of the sport may have thought the ideal racing pigeon should look like physically, the survival of the fittest to perform the tasks imposed has moulded the physical type of the present-day racing pigeon. Bearing this in mind, and while observing in the plumage, colour, markings, nesting and feeding habits how closely both the racing pigeon and the feral street pigeon resemble the Rock Dove, any slight difference can be accounted for by environment. This similarity sets me wondering: are these feral pigeons descendants of the Rock Dove which, finding food and suitable nesting places in our towns and cities, have migrated there? Or are they the descendants of domesticated varieties which, for one reason or another, have been forced to fend for themselves, and from random matings have reverted to the type of their common ancestors? Equally, of course, they may be the product of both: who knows?

We know that many of the fancy varieties could not survive without the care and protection of man. Others, moreover, if left to fend for themselves, would have a precarious time in the struggle to survive. However, by mating with other varieties their progeny become more fitted to survive in their new environment than the parents, thus preserving the species if not the variety. Irrespective of how much a variety may come to differ from the Rock Dove in appearance and from environment in manner of life, they all possess the same inherent instincts of the species. They are monogamous, yet gregarious and unique in two respects: first, in their manner of drinking and secondly in that both sexes produce in their crop pigeon milk or soft food, with which they feed their young, by regurgitation.

Writers at the beginning of the present century, more conscious than we are of the cross-bred origin of the racing pigeon, warned their readers of the dangers of reversion. They saw in the sudden and unexpected appearance of a character in the young, not possessed by either parent, yet known to have been possessed by a remote ancestor, symptoms of degeneration. In this respect racing pigeon fanciers have an advantage over the breeders of other livestock in that any errors of judgement in selection are sooner or later corrected by the test of the basket in training and racing. Pigeon racing is a strenuous sport and only the fittest successfully survive: this applies to the man as well as to his pigeons. I have been told in the strictest confidence many secrets concerning racing pigeons and their management, but declare there are no secrets in the way of magic formulas which bring success. Common sense, observation and a determination to get at the hard facts of every problem with which the sport abounds will fascinate and hold you for a lifetime.

The simplest way to understand the behaviour of racing pigeons, together with the probable source of the stimulus which brings into action each activity observed, is to try to watch the gradual development from the first day a youngster is hatched. The newly hatched young pigeon is indeed a helpless creature, seemingly only able to waggle its head.

All it then requires is warmth and food, as provided by its parents.

Self-Preservation

The young pigeon can do nothing towards seeking either, except that this head-wagging acts as a stimulus to the parents to feed. After several days it will be noted that it can see, and later still, if neglected, it will call to be fed, or if you care to put your hand or other strange object near, it will raise itself on its legs and pick at the object with a clucking noise. This is the manifestation of perhaps the most primitive and strongest instinct possessed by every living creature; self-preservation. At this time, this reaction to fear is all the young pigeon is capable of in self-defence.

Company and Appetite

When young pigeons are weaned by being taken from the nest and placed in the young bird compartment or loft, almost at once we see evidence of another instinctive characteristic of pigeons—their liking for the company of other pigeons. Newly weaned youngsters will cluster together, usually in the farthest corner: in addition to the attraction of the other youngsters, perhaps they are attracted by the warmth. Left alone, they soon settle and the stimulus of appetite then urges the most forward to seek food. Their first impulse when hungry is to seek it where they last found it and that is from their parents. It will usually be found that the first young of the season are just that bit longer in learning to pick and eat for themselves than the later lots. Once one or two start picking tentatively at the grain put down for them the others soon follow their example. Just watch how these younger birds at first keep trying to put their beaks in the mouths of the others, squeaking and flapping their wings. These actions have an appetite-stimulating effect and excite the others as well as themselves.

In the majority of racing pigeon lofts the weaning of youngsters is a sudden affair. One minute they are in the nest, the only place with which they are familiar, and the next they are in the young bird compartment, which to them is another world. This sudden change of environment must be somewhat bewildering, as must be the need to seek and eat their food which so far has been provided for them. To offset this, many fanciers put a dish of food in the nest box so that the youngsters, seeing their parents pick, will learn to do so. The advantages and disadvantages of both systems will be discussed fully later. Under either system the weaning of racing pigeons is an artificial affair, brought about in most instances for convenience in the management of the parents as racers, since the sooner the youngsters can fend for themselves the better.

Feral Weaning

With feral pigeons their young are restricted to the nest and its immediate vicinity until they can leave it by flying. When the parent birds go to nest again the young are gradually neglected and can be seen and heard squeaking and chasing after their parents to be fed. With this treatment they very soon lose their plump baby flesh and take more readily to the wing and follow their parents to the feeding place and learn to search for food and to eat. By this time they are not the plump heavy youngsters of the same age found in the racing pigeon lofts, but lean, wiry and wary pigeons. The weaklings have gone under in the struggle of the survival of the fittest.

Fear and Coddling

Beginners and others continually fear that their pigeons, young and old alike, may suffer a set-back, and this drives them willy-nilly to coddling their pigeons, particularly in feeding. The actions of these fanciers seem logical enough, and having taken the youngsters away from their parents they go to great pains to see that they get plenty to eat. At the end of the day they will go carefully over these newly-weaned youngsters; and those which have not eaten anything or, in the opinion of their owner, not enough, they proceed to pack with food. I, too, have packed such youngsters, spending hours at it, fearing they would come to harm. I have even gone to the trouble of steeping the beans or peas in water until they were swollen and then, before giving them to the young, putting the beans or peas into hot water to take off the chill. Now I realize that it was a lot of unnecessary trouble, and entirely illogical. I am referring only to youngsters that have been removed from the nest to be weaned. Hunger is the stimulus to appetite, and when you pack a youngster its hunger is satisfied and its appetite, or stimulus, which would otherwise have driven it to seek food and learn to eat, has gone. I have found that if you have packed a youngster once or twice, it has a tendency to rely on you to feed it: instead of helping, you have actually prolonged the process of teaching it to eat for itself.

Logical and Practical Weaning

When faced with the backward youngster the one thing I make sure of is that it has taken a drink of water, either of its own accord or by my forcing water over its throat with a rubber ball syringe. Alternatively, I may push two or three beans over its throat. The reaction here is surprising. Many a time a young bird which had shown no interest in food, on getting some water in the manner described and a couple of beans, has reacted immediately by picking at the food offered it. More often, to stimulate their appetite, I wait until my old birds have been fed. I then put the backward youngster back into its next box and allow one of the parents to regurgitate a little water and food, as a stimulus to the youngster's appetite. To allow the parent to feed the youngster to any extent would be sheer folly.

Familiar *v.* Unfamiliar

Once youngsters have learnt to eat and drink, they very soon begin to fly up into the perches and take stock of their immediate surroundings. This is their home, the familiar place in which they find the things they desire: food, water, the security of their perch and the company of their kind. I find that it is wise to get youngsters out on to the loft top or trap at the earliest possible moment, even before they have learned to fly up into the perches. They will usually try to re-enter the loft at once if we allow them to: the fact that they do so is natural and desirable. Inside the loft is their familiar abode and the big outside world is strange and unfamiliar to them. Gradually they will stay out longer, surveying their surroundings and, perhaps, the most venturesome will take a short flight themselves. At this stage it is of the utmost importance that we have some control over these youngsters. They have come to lose some of their original fear of the world outside the loft, which in the first instance was what drove them back to the security of the loft. Once they are outside, however, if we have not taken steps to have some sort of control over them, we are entirely at their mercy as to when they decide to come back into the loft. They may sit outside for hours, and only when they begin to feel hungry or thirsty or, with the approach of night desire the security of their perch, will they come in.

Natural Inclinations

Young pigeons which are habitually allowed to sit about the loft or roof tops and re-enter the loft only when they feel like it, are making a bad start in life for which their owners are to blame and will assuredly pay dearly. The golden rule is never to let young pigeons out unless you know they are definitely hungry. Even one with some food in its crop from a previous meal can upset and spoil others, wasting our time waiting for it to enter. These youngsters which have become accustomed to the outside of their loft are motivated by several instinctive desires, feelings and emotions. These may operate singly or collectively and the individual pigeon acts on the most pressing desire of the moment. The strongest and most instantaneous in action is that of fear or self-preservation. Pigeons react to this in several ways: the most common is to take to instant flight, another is to freeze or, in defence of their nest, to fight.

Just watch some youngsters on the loft top that take fright: those that can fly, fly off into the air at once while those that cannot fly, freeze momentarily and then scuttle back into the loft. We also notice that these young birds are exceedingly gregarious, staying together, following each other around in a group and if one takes to the wing the others usually follow. Sometimes, in the wild scramble of taking to the wing for the first time, one will alight on some distant building, but sooner or later it will come back among the others.

In summary, we find our birds are motivated in their actions by fear of self-preservation, the company of others, the security of their perch, food and water. There is not much their owners can do about the first three, but we have complete control over their food and water and by their intelligent application we can obtain a surprising measure of control over our pigeons. This subject will be dealt with fully in a subsequent chapter.

The Empty Loft

We see, then, that the reason why pigeons return to their loft is that all their desires and interests are centred there. I have had the experience of a loft of pigeons being sold and the loft left standing: over and over again birds returned to this empty loft but, being given neither food nor water, they took their departure in two or three days. It was a different matter if two birds of opposite sex came at the same time; even if not given food or water they would stay around, fending elsewhere for food but returning to the old loft at night, presumably for the security it offered them. These facts have been confirmed many times by others. Similarly, if a bird returns to its old loft and is immediately put in a basket away from the loft and given only water, it will usually clear off, on being released in reasonable weather, if not at the first attempt then at a later try.

The pigeon that returns to its old loft does so in order to satisfy some urge or desire seemingly not found in the new loft. Therefore, if food is denied it in the old loft, the chances are that it will return to the new, where it last found food.

Memory and Food

This characteristic trait of pigeons of being able to fix a location in their memory, and particularly if it is in any way connected with food and they are hungry, must never be underrated. The following incident will perhaps show the reader the acuteness of this faculty in pigeons. Once a team of pigeons is trained on the 'Boomerang' system, it is a comparatively simple matter to change the location of the point to which you desire them to fly on being liberated from their loft. The method is, you put the team of pigeons into a basket and then take them to the point to which they are to fly. On reaching this point make a show of preparing to feed them in the basket and then chase them out into the air. As they circle you call them back into the basket and give them a little to eat. You keep repeating this, each time forcing them to fly longer and further from the basket as they circle. When you feel they have had enough to eat, give them a drink and remove the basket out of sight, when usually they will return right away to their loft. After three or four days of this treatment you can confidently release them from the loft and they will fly out to the selected point and enter the basket to be fed.

I recall vividly one evening a very crestfallen young man reporting to me that he had lost every one of his team of 'Boomerang' pigeons. On my instructions he had taken these pigeons to a point some fifteen miles north-east of their loft and went through the procedure detailed above. Everything proceeded normally; he removed the basket and the birds cleared away. On his return to the loft he received a shock to learn that none of his birds had returned. He immediately thought that they had gone to the point to which they had been previously flying which was some ten miles south-east of the loft, but although he waited there until it was dark they did not put in an appearance. That night the weather became bad, with mist and rain, and next morning no pigeons were at the loft. What puzzled me was that not one had returned to the loft. Being asked to fly fifteen miles in a new direction might account for a pigeon or two being lost or arriving late, but here were a dozen or so gone.

As usual when things like this happened it was the owner who had blundered. It had proved to be rather late to liberate pigeons from this distance for the first time, and while the weather at the liberation point was fair, at the home end it had deteriorated rapidly. As no pigeons had arrived at the loft I reasoned that after spending a night out they would be hungry and looking for food, so I dispatched a man to the old point ten miles south-east and my crestfallen young man back to the point to the north-east where he had liberated his pigeons. In a couple of hours they were both back. The majority of the pigeons were found picking around the exact point where they had been liberated and were all so wet that they could hardly fly and quite happily walked into the basket. Two or three were picked up at the old point and, if I remember rightly, one turned up at the loft a day or so later, when the weather had cleared. I fully expected that most of these birds would be picked up at the old feeding place to which they had been flying for a week or two, but it just proves how little we know about pigeons and their reactions in any given situation. I would have liked to have known where they spent the night, and if they were scattered or all together. From personal observation of the behaviour of pigeons I am doubtful if the reason for their return to their lofts is their sheer love of home.

They Live Where their Interests Lie

This is something of a myth for, if you understand your pigeons, it is easy to settle one very quickly to a new loft and, strangely enough, the better the bird the easier it usually is to settle.

Recently I received a pigeon the day after it had flown 500 miles and within two weeks I had it settled and flying out and in, as if it had never known another loft. It is true that the first time he got out, after the sexes had been separated, he went off to his old loft. When noticed there, where the sexes were also separated, he was put in a basket and removed from the loft. Next day he was liberated and found a closed loft, so off he went back to the new loft where he was allowed to see his hen for an hour or two, before being separated again. The old adage 'you never miss the water until the well runs dry', applies very much to pigeons, and its application in our management of them can be used to our advantage.

Stratagems

Take away a driving cock from his hen and he becomes keener than ever, or, for that matter, a bird from its eggs when it should be sitting. There are many other similar situations in which stratagems can whip up the intensity of natural desires, feelings and emotions. They must not be used indiscriminately and not too often on the same pigeons. While they may not reason in the true sense of the word, they learn very quickly from previous experience and while they may react in the first instance in some particular way to a given situation, later on they may behave very differently.

Many years ago I had two cocks, a Dark Chequer 349 and a Blue 374, who were rivals in every way. As yearlings they both twice won prizes; as two-year-olds, I just could not make up my mind which was the better pigeon. Out of six races in which they both competed the Blue beat the Dark Chequer three times and the Dark Chequer his rival twice. In the other race, I timed them both in the same thimble to win 1st and 2nd Club. In the final inland race I sent them to different race points with different clubs with the result that the Dark Chequer 349 won 3rd LWC Weymouth 361 miles and the Blue 374 5th Lanarkshire Social Circle Salisbury 337 miles. Incidentally, they had both been my first bird from these race points the previous year and 349 was timed on both occasions. I now set about preparing them for a 500-mile open race. Just at this time I had a visit from the late Dr. W. Anderson who, like many others, had been caught up in the 'Eyesign' epidemic which was then sweeping through the fancy. Naturally, the subject was discussed and demonstrated and I took the opportunity of asking him to say which of these two cocks was the better pigeon. He selected the Blue 374. A few days before they were due to be basketed I was astonished to find these two rivals courting each other instead of fighting and I became really perturbed when I caught them actually treading each other. I wondered what this meant in terms of their fitness to race, for they looked and behaved like fit pigeons.

They were sent to the race, 374 sitting ten days on his third nest and 349 hatching; both were growing their third flight. I nominated 374 and as is usual I timed the Dark Chequer 349 first at 8.18 p.m. However, I was barely out of the loft when 374 arrived and was timed at 8.27 p.m., both in the prize list.

I have thought a lot about the reason for the change in the behaviour of these two rivals, and have concluded that gradually they became reconciled to each other's presence, which now cease to arouse the natural aggressiveness of one cock to another. Perhaps if they had been driving or feeding young it would have been different; as it was, they each could see their mates sitting quietly on their nest. Being very fit they apparently found an outlet for their surplus energy in courtship with each other.

Nerve Impulses

In very fit birds more nerve impulses are generated than are always required and these have to find an outlet as a sort of safety valve. One of the signs of a very fit pigeon is its apparently unblinking eye or, especially with hens, their trembling tails and wing tips. I am of the opinion that both these phenomena are outlets for their surplus nerve impulses. I do know that with the very fit pigeon the reason for the apparently unblinking eye is that the third eyelid at this time blinks so rapidly that it is hardly discernible even when watched for.

There are many characteristic acts of behaviour, the causes of which we have not yet fully understood.

Time Sense

There is any amount of evidence that pigeons have a very fine physiological time sense: for instance, the regular average hours at which the first and second eggs are laid and the hours at which the sexes take over their turn at sitting on the eggs. The position of the sun must play an important part in this timing, as I am convinced it also does in the regulation of the moult.

The Sun

In Great Britain the unmated pigeon or the very late moulter will normally have cast its first flight by Midsummer Day, and it is unusual for a pigeon to throw any more flights after about the 22 December. In the mid-twenties I sent some birds out to South Africa in October, after they had completed their body moult. They had barely settled down in their new abode when they started to moult all over again. The reason was that while they left Great Britain at the beginning of the winter they arrived in South Africa at the start of the summer there, and apparently they immediately fell into step with the season of the year in which they found themselves.

Moulting

There are a number of other factors affecting the course of the moult. It is generally accepted by racing pigeon fanciers that the date on which a pigeon is mated controls the fall of the first primary flight and, therefore, the others in their sequence. Many fanciers delay mating up, with the idea that by doing so they can send their pigeons to the final long-distance races growing their second primary instead of the third, or their third instead of their fourth. My experience is that there is a limit to the number of days you can delay the natural moult without running into trouble. Pigeons for which the moult has been delayed or suspended, either by being mated up late or because they were sick or injured, on starting to moult again are more than likely to cast two instead of one flight in each wing. My observations and data show that the majority of racing pigeons cast their first primary flight when sitting on their second round of eggs. Normally the flight is cast on or before the tenth day of sitting. There are a number of pigeons which do not cast their first flight until sitting on their third round of eggs. These late moulters appear to have inherited this trait and pass it on to a proportion of their young, and you will find even nestmates, one a normal second nest moulter and the other not casting its first primary until sitting on the third round of eggs. My attention was first drawn to this by the difference in moulting of my 'Nantes Pair'. This pair were mated together for the four years in which they both were sent to Nantes and each year the cock 2030 was basketed having cast his fourth primary flight, while the hen 2033 would just have cast or be growing her second primary flight at basketing around the 15 July each year. The undernoted are the dates on which they each cast their first primary flight.

	1st Year	*2nd*	*3rd*	*4th*	*5th*	*6th*	*7th*
2030.	6 May	12 May	10 May	18 May	20 May	18 May	18 May
2033.	16 June	20 June	20 June	23 June	20 June	20 June	20 June

It has been suggested that late-bred birds hold their flights longer than early-bred birds, but in my own experience I have found no evidence to substantiate this. It is true that late-breds sometimes do not complete the moulting of their primary flights in the year of their birth, but in successive years they moult normally according to their inheritance.

The following are some examples. The first is that of a son and daughters of the Nantes Pair, all of whom have flown 500 miles.

	Yearlings	*2nd Year*	*3rd Year*	*4th Year*
2089 Hatched 31 March	25 May	6 June	28 May	1 June
2104 Hatched 25 July	25 May	6 June	5 June	7 June
2112 Hatched 19 May	2 June	5 June	5 June	7 June

For comparison, this second lot are four brothers and sisters, 500 miles, hatched at different dates.

	Yearlings	*2nd Year*	*3rd Year*	*4th Year*	*5th Year*
2109 Hatched 11 May	26 May	6 June	1 June		
2102 Hatched 16 June	25 May	6 June	5 June	8 June	
379 Hatched 24 June	28 May	15 May	27 May		
392 Hatched 28 July	27 April	18 May	25 May	25 May	18 May

The undernoted list gives the dates of the first primary casting of a group of pigeons which raced up to their fourth year and flew 500 miles at least twice.

	Yearlings	*2nd Year*	*3rd Year*	*4th Year*
349	7 May	15 May	15 May	17 May
547	4 May	7 May	10 May	18 May
2041	3 May	5 May	9 May	18 May
3838	25 May	4 June	1 June	5 June
2843	3 May	17 May	20 May	18 May
2084	15 May	7 May	25 May	25 May
6243	7 May	17 May	1 June	27 May
6248	12 May	28 May	26 May	1 June
6258	12 May	17 May	5 May	19 May
6267	27 May	10 June	17 May	25 May
6276	20 May	25 May	10 May	27 May
6271	12 May	15 May	15 May	15 May
6273	12 May	12 May	25 May	30 May

It will be noted that the majority of the birds in the above lists cast their first primary a bit earlier as yearlings than in later years. This is accounted for by the fact that as yearlings they were mated a week to ten days earlier than in the years following.

After the yearling stage, the casting of their flights is very regular, except that 6258 cast his first flight as a two-year-old on the 17 May, as a three-year-old on the 5 May, and as a four-year-old back to the 19 May. On checking this I found that he reared only a single youngster in the first nest and his mate laid again rather quickly her second round of eggs, and I feel sure this accounted for 6258 casting his flight on the 5 May. My experience is that it is exceptional in normal circumstance to find a pigeon cast a flight other than during the period between the laying of the second egg and the tenth day of sitting. There seems to be a lull in the procreative processes at this time during which the fall of the flights takes place, caused perhaps by activity of the thyroid. By the twelfth day of sitting the formation of the soft food has begun and from then until the young are reared the casting of feathers is suspended. The feathers which have been cast are regrown: I have witnessed this many times, where a pigeon has completely regrown, say, its second primary flight but the third primary did not fall until the next round of eggs was laid. I called to see a winner of the Scottish National the day after the race and while I was handling it a flight dropped out of one wing and the corresponding feather in the other wing was loose and ready to fall. This pigeon was sent to the race with a big youngster in the nest box and his mate had laid her first egg the day previous to basketing, so that the fall of the flight happened on the sixth day of incubation. Summarizing the situation, under normal circumstances I consider that:

(1) The date on which a pigeon is mated and the first round of eggs are laid influences the fall of the first primary flight.
(2) Pigeons cast their first primary flight while sitting on their first, second or third round of eggs according to whether they are by inheritance, early, intermediate or late moulters. The majority are intermediate moulters casting their first primary flight when sitting on their second round of eggs.
(3) The period during which flights are cast occurs between the laying of the eggs and the tenth to twelfth day of sitting.
(4) The period during which the casting of feathers is suspended occurs from the formation of the soft food and continues during the feeding of the youngsters until the next round of eggs are laid. Injury, sickness or any serious upset from whatever cause, such as extreme exhaustion (being flown down), suspends the casting of feathers until the bird concerned returns to normal health.
(5) Unmated pigeons cast their flights according to their inheritance; while they may cast their first flight slightly later than birds which have been mated they soon make up for lost time.
(6) Pigeons which have had their state of moult retarded by the suspension in casting from whatever cause will, on return to normal, cast the extra feathers, thus, as it were, bringing their state of moult up to date.

Before leaving this very interesting subject there are one or two observations I have made that are worth recording. The first is that the observant will have noted that the early indication that the mating season is at hand is the appearance of small down feathers in the quiet backwaters of the loft, or sticking to rough surfaces on the walls or partitions. This is the under down. If you examine the thick down around the root of the tail you will feel the empty sheaths from which this down has been discarded. Yearlings are much later in reaching this stage than older pigeons and I think it is for this reason that yearlings do not come to that tightness of body feathers until mid-June and into July, a period in which they put up their best performances of their first season's racing.

The second point is that it will be noticed that the first signs, apart from the casting of their first primary flight, that youngsters have started moulting is the growing of two rows of small feathers across the middle of their wings. These feathers are easily seen as they are richer in colour than the other juvenile (nest) feathers above and below them. Although it would appear that these are the first small feathers to be cast and regrown, I have watched this carefully and have come to the conclusion that these two rows of feathers were never grown as nest feathers. If you care to examine the wing of a feathering youngster just before the pen or pin feathers burst you will find that the tiny morsel of flesh and bone which is the wing proper is a mass of pen or pin feathers. Just notice that while the wing proper is comparatively small the pens are not proportionately miniatures, but full-sized. In consequence of this there is a limit to the number of pen feathers the wing can hold until the flesh and bone of the wing proper has fully developed and it is only then that these two rows are grown, for the first time.

There are those who hold the belief that pigeons suffer pain or at least discomfort when the new pen is at the stage of bursting through and therefore it is asking for trouble to send a pigeon racing at such a time. This is a matter of speculation based on the assumption that the growing feather emerging from the follicle is a painful process. When I think of this situation, it makes me laugh; just think of the agonies a young bird must go through while growing hundreds of feathers during the first few weeks of its life. The above is just one of the many plausible hypotheses believed and passed on without any attempt to prove or disprove it, and makes pigeon racing more complex than it really is.

Finally, beware of exaggerated features as displayed so clearly in the broad-flighted, over-feathered show specimens with their immoderate heavy-as-lead bodies, which obviously could not fly to save their lives. Yet some judges must be awarding these monstrosities tickets otherwise they would not be exhibited. I declare that there is no great difference between a nice racer and a top class show-bred bird, but I must admit that the show bird often gets the better of the genuine racer by the overall finish which is bred into it. When judging, I have been appalled at the number of the exhibits I need never have taken out of their pens. Some years ago, when judging a class of over 100 pigeons, I became alarmed when I had reached the twenty-fifth pen without marking a single one: fortunately there were pigeons in the class to which I had no qualms in awarding tickets. A pigeon may satisfy a judge's fads and fancies when critically examined by sense of touch, be free of blemish of feathering and spotlessly clean, but it must look the part of a winner in the pen judged by sight alone.

Outstanding Characteristics

The above chapter goes thoroughly into the stimuli which motivate the actions of pigeons in given circumstances, but let me reiterate briefly a few of their outstanding characteristics. First and foremost their manner of drinking, which consists of placing the mouth into the water and sucking it up into the crop. In around seven to ten days from a pair taking to each other a single egg is laid with a second one some forty-four hours later. Both parents share the incubation of the eggs and the feeding of their young. This is done by regurgitating the soft food produced in the parents' crops which is gradually reduced until some ten days later the parents regurgitate the fresh food they have recently eaten.

2 Environment

Racing pigeons are simple creatures without guile, motivated in their actions by the desire of the moment, and because of the environment we have created for them they are able to satisfy most of their desires in the loft. Their environment is the air they breathe and the food, drink, exercise and training they receive, in short, everything which affects their mode of life. It is evident that while the inherent instinctive urges, feelings and emotions are basically the same in all pigeons, each loft has its own peculiar environment created by the individuality of the respective owner. Visit a number of lofts and you will quickly come to recognize these differences; moreover, once you begin to realize what causes them, you will learn a lot about the owner of a loft from the behaviour of his pigeons. While environment cannot change the basic instincts of the species it can and does modify them within certain limits. It can make the naturally tame pigeon tamer or the wild pigeon less wild or vice versa; spoil the good pigeon or make it better still.

The Effects of Environment

The influence of environment on the future character of a pigeon begins from the moment of fertilization. It consists of the quality and quantity of the contents of the egg and, of course, the warmth provided by the parents during incubation. After hatching it consists of the warmth and soft food provided by the parents and the general attention received until the bird is weaned. While at Old Neilsland, I noticed that one or two youngsters were, when first weaned, just that bit wild or perhaps more timid than the others, and at first put it down to an inherited trait. When it happened again and the parents of these youngsters were among the tamest in the loft, I started to think again. Eventually I stumbled on the solution. The nest boxes in which these timid youngsters were reared were high up in the roof of the loft. To get them I had to bring in a box to stand on and because of this I saw very little of these youngsters or they of me or what was going on in the loft. The youngsters reared in the other next boxes could see all that was going on, and I recall the children on their occasional visits to the loft were delighted at being allowed to touch these youngsters.

Thus, even at this early age these youngsters were learning not to be afraid; at least they did not panic like those reared in the out-of-the-way nest boxes did when they were first weaned. A similar example is to be found with single reared youngsters which, in spite of a pigeon's natural gregarious instinct, will, when first weaned, resent the approach of other youngsters. One of the most interesting cases in my experience of the effects of early environment on the future character of a pigeon is that of 'The Outlaw'. His sire, a prisoner, escaped and was never seen again and 'The Outlaw' was reared singly by his dam. When he was about sixteen days old his dam took up with the cock in the next nest box. This cock was a quiet enough pigeon but when he entered the nest box to be with the hen he pecked at the big youngster, not viciously but really in self-defence to keep the youngster away. 'The Outlaw' reacted by flapping wings and squeaking and usually succeeded in chasing the cock out of the nest box. The result of this was that when he came to be weaned he would not allow another youngster near him. He fought for and claimed the highest and most remote perch in the Young Bird loft and in fact he was a perfect menace. Twice he disappeared for several weeks, eventually coming back on his own, but because of these absences I did not have much chance to tame him. Although as a young bird he was twice my first bird and never out of the first half-dozen flying 220 miles and during his five racing seasons he won many prizes, he remained antisocial and intractable. These are just little pointers to the effect of environment on the future character and behaviour of every pigeon and evidence enough of the desirability of a carefully-thought-out system of training right from the day they are hatched.

We know that pigeons like what they have been used to; in other words, they like the environment in which they have been reared and have found acceptable. Let us consider the impact on pigeons suddenly exposed to a strange environment. I have seen young pigeons having to be chased out of a training basket, because they fear the unknown. I have even seen them run back into the basket, probably because they have been taught to eat and sleep in the basket when being weaned. I know that many fanciers carry out this practice but personally I do not approve. The only place I want my pigeons to long for is the inside of the loft. It is popularly thought that what brings a pigeon back is its great love of home but, as I pointed out earlier, this is not strictly correct. It is rather a shattering thought that it could be the reverse and that what, in fact, impels our pigeons home is fear.

Is Fear the Driving Force?

An unknown and strange environment into which they have been liberated frightens them and they race home to their loft and the known security it affords. This thought will not go down well with the sentimentalists but it is well worth considering. We have all known the pigeons which, as young birds, yearlings and two-year-olds, have raced very well indeed but we have later regarded as cute because they have ceased to hurry home. We know that pigeons learn from experience. Could it be that what has happened to these older experienced pigeons is that they no longer fear the unknown or at least fear it less and in consequence they no longer panic?

I have spent many an anxious hour because a good pigeon has failed to return with the others from a training toss or short race. I have thought that their lateness was the result of their independence of action, since this kind of bird usually turns up trumps when the going is tough. Cool, calm and confident they set their own pace which enables them to outlast their more impetuous rivals. However, I think it is something worth considering. Just watch the returned youngster or yearling, observe its demeanour as it alights on the trap and ask yourself whether it does not look like a frightened pigeon. This is particularly so with the early arrivals which consequently have not been flying with the big batches. We send a team of pigeons racing and at the end of the day we try to sum up the results. Why are some early, others late and why have some failed to return? Sometimes we think we have the answers, but in the main we can only speculate. Racing pigeons are exposed to many hazards: telegraph wires, power cables, radio and television aerials, shooting, trapping and their natural enemies, such as hawks and cats. Only an infinitesimal number of the victims of these hazards are ever heard of, and many of these never by their owners. In a lifetime's experience I have come to the conclusion that only a comparatively small proportion of the pigeons lost enter other lofts. What happens to them, where they do go, are ever-present questions in our minds. Speculation as to their fates could fill a book. We know that they are picked up by ships at sea, but for every one heard of there must be dozens who disappear completely. I recall a pigeon with a good racing record being picked up drowned in a reservoir 450 miles from the race-point and 100 miles from its loft on the day after liberation. Did it fall into the water trying to get a drink, or was it chased down by a hawk? Its fate would never have been known but for the fact that it was seen and picked up by a local fancier who reported it. I have seen a racer come down and take a drink of sea water which I should not imagine would help its thirst. We have all had pigeons fall down chimneys and I hate to think of how many returning racers overtaken by approaching night have met this fate in attempting to alight. We hear of isolated incidents and we need only think of how often they can probably be repeated all over the country to get an idea of what is going on. I once had a young bird reported, and called at the finder's loft to get it. On arrival I found that his birds were exercising and mine was with them. I was told that a few had alighted but the others had gone off to the fields and mine with them. Looking at the birds on the loft top I pointed to one and was told that it also had an M ring. I said that it was one of mine. We got this bird into the loft and sure enough it was mine. I took this one away and told him to keep the other. Both these youngsters had been lost from my loft but not at the same time nor did they arrive at their new loft at the same time. Why, then, should they both reach the same destination? Incidentally, the youngster I took back with me turned out to be an excellent racer.

Liberations

Pigeons are by nature gregarious and from their environment racing pigeons are particularly so. They habitually exercise around their loft as a team without a leader, their evolutions apparently controlled by the will of the majority as they turn and swing around in the vicinity of their loft. On being liberated from a basket these leaderless evolutions as they circle are repeated in every detail, particularly so at the shorter distances. When a convoy of pigeons is liberated at the longer distances they normally clear off right away, but why this is so leaves room for discussion. From the point of view of environment this has changed considerably compared with the earlier shorter races,with their preponderance of inexperienced, poor or less fit birds composing the convoy. As the distance increases it is reasonable to assume that the convoys are composed of birds having some experience and that as a whole they are all better fitted physically, by experience, for the task in front of them. A study of race results suggests that as the race proceeds a vanguard is formed of groups of, presumably keen, fit pigeons which gradually draw away from the main batch. As the hours on the wing increase these batches thin down and break up into smaller groups. I have watched such groups go past when the birds in them had still 100 miles or more to do before regaining their respective lofts. I have watched a dozen or twenty birds go past with the leaders going strong, while one or two birds were trailing behind losing ground and obviously finding the pace set too much for them after nine or ten hours on the wing. Everyone will have watched the breaking up of batches nearer home; you will have noticed one pigeon pull away to one side and the others follow it for a little, only to swing away to their original course taking the single bird with them. You might even see this bird finally break away or even turn back: evidence of the desire for the company of others. We have all experienced a batch going over and when well past a bird breaking back to our own loft. This gregarious instinct of pigeons influences the results of races at all distances. If the majority of pigeons in a batch are for one particular area they will dominate the others to the extent of pulling them away from what would otherwise be their direct course. Disgruntled fanciers in their disappointment and disgust declare it is all a matter of luck. This is only partially true; there is the luck of the wind and the weather, which is all part of the game and must be accepted as such. There is also the element of chance that your pigeon is in a good batch from the start, and that this batch gets the breaks our changeable weather creates *en route*. But to win, a pigeon must have the urge and physical ability to last the pace for the hours required to complete the journey, without which no amount of luck will avail.

Hours on the Wing Which Kill

The measure of performance of pigeons is given in miles flown, but my experience is that it is the pace and hours on the wing which finds them out. Study the results of a few very fast races of over 300 miles or more and watch how quickly the velocities of the winning pigeons drop as pace takes its toll. Study any race where the winners are taking more than ten hours to do the journey and as a rule there will be empty perches at nightfall. It is pace and hours on the wing which bring exhaustion, and we will now consider what effect this has on the desires of the pigeons concerned as they plod on hour after hour. Physical exhaustion brings about an overwhelming desire for rest, so the fatigued pigeon stops flying. Furthermore, if my observations are correct, the next most desirable thing after rest will be water to drink. If the bird has found somewhere to rest and water to drink, a night's rest and the body's reserves will have replaced some of the losses, and the pigeon may resume its journey the following day, without having eaten any food. This raises the question of whether we prepare and condition our pigeons correctly to meet the task we set them, especially for the longer distances which perhaps involve a night out.

Preparing a Record Breaker

In 1927, the pigeon world was electrified by the performance of a Blue Chequer Hen NU22YAX.6 and R.P.6 in flying 788 miles at a velocity of 939 YPM. This was a record for the distance which the late Lt.-Col. A. H. Osman thought would stand for a long time. Now let us see how this record-breaking pigeon was prepared for the great day. As a youngster she was trained and won 1st Essendine and Huntingdon 120 miles. As a Yearling in 1923 she was described by her owner as being a sweet beautiful hen. She was not raced but used as a hen for a widowhood cock, and when not required for the odd hour or two with the cock was kept with a number of other widowhood hens in a cage measuring 3 × 3 × 8 feet and was fed only once per day.

As a two-year-old in 1924 she was treated in the same way until later in the season, her mate having been lost, 'No. 6', without a moment's exercise, was taken from the cage along with three other hens and jumped into a race from Bournemouth, 242 miles. She not only beat the others by two hours but won the local club race. After this race she was given a mate and allowed out once per day and although due to lay when basketed she was sent with sixteen other birds to a race from Paris.

The race was a hard one and of the seventeen sent 'No. 6' was the only bird home on the day, winning 1st Club, 3rd Yorkshire Mid-week and 3rd East Sect, Great Yorkshire Amalgamation. She was then sent to Nevers, 520 miles, again due to lay her second egg in the basket, but nevertheless, she won 1st Club, 3rd Yorkshire Mid-week and 5th East Sect, G.Y.A. and all pools. Her owner tells us that the work done by 'No. 6' in 1924 taught him two lessons. One that a big jump is beneficial and the other that very little food is necessary to condition a pigeon. When the 1925 season came along 'No. 6' had not moulted well and looked anything but fit, but thinking a big jump would do her good her owner sent her straight from the loft to Folkestone, 216 miles. On her return she seemed to 'bubble up' and so was sent to Pont L'Eveque, 332 miles. She did this on the day but was obviously out of condition and was stopped for the rest of the season. Having moulted well when 1926 came along 'No. 6' looked more like her old self and was jumped to Folkestone. Standing in the loft some twenty minutes after timing his first bird, her owner was startled when a pigeon tumbled in the loft door like a stone at his feet. Picking it up he found it was 'No. 6' with her feet stuck by dried blood to her feathers so that she could not stand. Both her legs were injured and two flights broken off in one wing. With her wounds dressed she recovered rapidly. Two flights were grafted into the stumps of her broken flights and a month later, although still lame, she was sent to Le Mans, 420 miles. In this race there were only twelve birds home on the day in the Middle Route Fed. and 'No. 6' was one of them, winning 1st in two Clubs and 6th Fed. In 1927 a race was being organized from Marseilles and it was decided to prepare and send 'No. 6' and this is what her owner has to say about her preparation:

' "No. 6" is a very small eater and inclined to put on fat as she will not exercise. When preparing her for the big test, she was made to go without food for twenty-four hours at a time. This I thought would help her on the long journey and keep her in the air. Testing my theory, on two days a week I took her from her nest and put her in a show pen in front of an open window and gave her neither food nor water. The following day I had her tossed three miles away and awaited her return. Each time on arrival she went straight to her nest and would remain there without food or water for hours. The idea behind this treatment was to get her accustomed to being without food or water for twenty-four hours at a stretch.' He went on to say, 'Cruel it may appear to some, I am convinced it served my pigeon in good stead, being of the opinion that she never dropped for food or water during her long 788 mile journey except during the dark hours.'

What manner of pigeon was this that stood up to all the trials and tribulations which formed the environment and experience of 'No. 6' now known as *Lady Marseilles*. Of her breeding her owner says: 'She is of the bluest of blue blood and stands today proudly asserting the victorious consummation of a strain distinct and apart from all other strains.'

Lady Marseilles

She was hatched on 23 March 1922, her sire being champion *Soranus*, a Blue Cheq. Cock 177–18, and her dam, *Princess*, a Blue Cheq. Hen HP17S–2345 and RP170–3405, bred by Jack T. Clark of Windermere from the well-known pair *Lakeland Regent* 3194 (Ref. C) and *The Gits* Hen 101 (Ref. E). *Lakeland Regent* was from *Lakeland Prince* 1926 (Ref. A) and *Miss Taylor* 34 (Ref. B), the latter pair being Jurions descendants of *Le Premier Revenu*. As to *Soranus* the published stud lists of J. T. Clark of 1921 and of F. E. Brown for 1923 both agree that *Soranus* was bred from 2170 and a daughter of *Lakeland Laddy*, 2170, bred by J. T. Clark in 1917 from 150, a son of *Lakeland Regent* and a Blue Hen 2030, a half sister to *Lakeland Laddy*, being from *Lakeland Blue Boy* 5970 and 578 a granddaughter of both *Lakeland Laddy* and *Lakeland Prince*. Both stud lists confirm that Mr. Brown had two daughters of *Lakeland Laddy* 8142, who incidentally was bred from *Lakeland Blue Boy* and a half sister to *Lakeland Prince*.

The first of these two daughters of *Lakeland Laddy* was a Blue, 1092, her dam being a Blue, 2681, winner of 1st Marennes 595 miles National F.C. Section A. from 9147, a sister to the sire of *Prince*. The other daughter was HP17S–2338 and RPO–3412, her dam being 1458 a granddaughter of *Prince*. These Clarks were a handsome family which not only won on the road at all distances, but held their own in the show-pen. For the record this Open race from Marseilles attracted forty-nine competitors who sent sixty-seven birds. Liberated at 10.35 a.m., 5 July, *Lady Marseilles* was timed just after 6 p.m. on 6 July. The next pigeon was timed by Harold Jarvis of Middlestown near Wakefield at 12.25 p.m. on 7 July, subsequently known as *Marseilles King*. Billy Pearson of Keighley timed the next two birds flying 803 miles.

Events of Note

I think it would be appropriate at this point to mention that in the years just prior to the First World War, when pigeon racing was beginning to emerge as a popular sport, a spirit of adventure was abroad in Great Britain. In 1913, a race was organized from Rome in which 106 birds were liberated at 4.15 a.m. on 29 June. The first bird was timed at 1.48 p.m. by C. H. Hudson of Derby on 29 July. This bird, *King of Rome*, flying 1,001 miles, created the record for the distance. The second bird, *Prince of Rome*, to Vester and Scurr of Spennymoor, homed on 18 August at 8.10 a.m., flying 1,093 miles. This was the longest distance flown into Britain and was not beaten until 1960. Thus it was that during my first years in the sport great things were happening. In 1910, the Scottish National flew to Rennes for the first time. This race was won by W. McLean of Dennyloanhead with his famous *No Surrender*. Mr. McLean emigrated to America and in 1960 returned for a holiday in this country and was an honoured guest at the annual dinner of the S.N.F.C., celebrating the jubilee of this race from Rennes. Then in 1914 several Scottish fanciers sent birds to the H.P. National from Bordeaux. Among them I knew Alex Black, John McIver and the Gillespie brothers, all of Airdrie, and Andrew Christie of Leslie, who found his bird in the loft on the second day but won nineteenth place in the race, vel. 739, flying 793 miles, a fitting climax to his having won the Scottish National the previous year. Incidentally, the Gillespies got a bird home in race time but without its rubber race ring. The late Tom Taylor of Blackpool was the first and second in this race, flying 630 miles. This will give the reader an idea of the calibre of my old friend Andrew Christie's pigeon, bought by Dr. W. Anderson when Andrew had to give up his pigeons on joining the Army Pigeon Service, (1914–1918).

Endurance

I have searched through the records of my lifelong friend

Jack T. Clark, which cover a period from 1896 to 1950, have checked my own records and have considered other sources, and I find that, as a general rule, if you do not get a pigeon back from a long race on the third full day, you are not likely to see it until a week or more has elapsed. This excludes pigeons which have evidently entered another loft and been cared for early on. I have concluded that the endurance of a pigeon is limited to three days of continuous effort and if my experience is any criterion there are very few that will carry on even for that time. In this connection I have noted that the majority of outstanding long-distance performances have been made by birds liberated late on the first day thus limiting their initial all-out effort. This is also true of races where the pigeons are put down by heavy rain before they become completely exhausted by physical effort for, when the rain clears, returns are surprisingly good.

The Importance of the Elements

An example of the importance of climatic conditions was the Scottish National Nantes race of 1956. After the birds were liberated the weather unexpectedly broke and it rained heavily and only ten pigeons were timed in race time. When the weather cleared returns, from all accounts, were beyond expectations. This race was won by John McGillivary and Son, Forth, vel. 690, who two years later won the Nantes race again with the record velocity of 1,478 for the 600 miles. Rain is always a good reason for not liberating pigeons and an explanation of low velocities and poor returns on the day. Yet provided the rain clouds are not low, resulting in poor visibility which usually denotes the absence of wind, the showers are intermittent, and the pigeons get a good start, they fly reasonably well in rain. Under these conditions there are usually considerable intervals between arrivals and, in consequence, irregular returns. On the other hand, if on a glorious summer day, with little or no wind, pigeons do not arrive when expected, the inexperienced wonder what can be wrong. It would appear that to remain airborne, pigeons need a wind to fly into. If there is no wind they have to create one by their own speed through the air, which demands continous effort on their part.

When flying, pigeons make use of every air current and if you watch them they can be seen gliding without a wing beat for considerable distances. In still air this ease of flight is denied them, as it is when a following wind exceeds their normal speed. In anti-cyclonic conditions, with combination of the light variable winds and heat, flying rapidly exhausts all but the very fittest birds if they are required to remain on the wing for ten hours or more. Paradoxically, there is a limit to the assistance a tail wind can be to pigeons when racing. Beyond certain wind speeds the pace pigeons have to maintain to remain airborne appears to create difficulties for them and therefore in extra fast races losses are much greater than in normal racing with velocities of between 900 and 1,250 YPM. Just watch a team of pigeons exercising in a strong wind and you will get some idea of what happens. Note how as they fly into the wind they gain height before turning to have the wind behind them. Observe how, immediately the wind is behind them, they lose height despite their rapidly increased wing beats. When racing with a strong tail wind they fly very high and in my opinion do so because they require height as a margin of safety should the attempt to glide in an overtaking wind fail. When we send a team of pigeons away racing we are inclined to assess their individual chances of success in terms of the previous distances flown. However, different weather conditions can make so much difference to each one of them, quite apart from the nesting conditions in which they were sent. All these things are part of the environment in which they may find themselves on being liberated and we can only conjecture how they will react if they have never experienced such an environment before.

New Environments

Reverting to the subject of pigeons brought to a standstill by sheer physical exhaustion; they may be able to find water to drink and a safe perch for the night, but what about food? By comparison with the feral pigeons the majority of lost and exhausted racing pigeons are ill-equipped by their normal environment to fend for themselves. They may search for food but in their strange surroundings they are not only unaware of where to look but of even what to eat. Their gregarious instinct may come to their rescue by their joining other pigeons at a farm-house or street pigeons in a town, and learning where to find food and to eat with them. Alternatively they may join up with some racing pigeons and enter another loft. In my experience, when an old racing pigeon joins feral pigeons and has apparently disappeared, it can be presumed that it is on its way home. With young pigeons, however, it is different. They are more adaptable, can adjust to the new conditions and may remain, whereas older birds tend to seek the environment to which they have long been accustomed. By their inexperience lost and exhausted racers fall easy victims to prowlers, both four- and two-footed. At a farm which I visited regularly there was an old dovecote with at least twenty pairs of pigeons nesting in it, and also half a dozen semi-wild cats living in the outhouses. Every season stray racers would arrive, spend a few days and then, provided they did not fall prey to the cats, clear off. From experience the regular inhabitants of the dovecote had become wary of the cats. Fanciers who, on the return of a pigeon that had obviously been in another loft, declare that it must have been kept a prisoner or who promptly kill it, anger me. They seem incapable of realizing what a state of exhaustion a pigeon may have reached before yielding to the overwhelming desire for food and shelter and the company of other pigeons. Have they never had other people's pigeons enter their own loft in a sorry state who refuse to leave until they are fit again? Much of this sort of talk is sheer bravado and in most instances can be taken with a pinch of salt. In fact, the majority of racing pigeon fanciers err on the side of being soft-hearted as far as their own and other people's pigeons are concerned. They create an environment in which their pigeons are coddled and then they wonder why they fail when put to the crucial test of racing from 500 miles.

Teaching Them What to Eat

What is the answer to this problem of preparing our pigeons to race from dawn until dusk and, if necessary, to spend a night out and be ready and able to carry on the next morning? I am inclined to think that there is a lot to be said for getting them accustomed to fasting for twenty-four hours, and can vouch for the fact that it does a fit pigeon no harm. Teaching them where to find food when they are beaten down and what they can eat is a more difficult problem. It might be helpful to teach them to eat the unhusked grains which they are likely to find in the fields, or bread crumbs, or all the other things they may be offered by those at whose house they may have settled. A recent experience illustrates what I mean. A pigeon was picked up so wet that it could not fly. It was evening and the lady who picked it up was at a loss to know what to do, so she put it on a shelf in the glass porch at her back door. Next day I received a telephone call from a newspaper office telling me about the lady and the pigeon. As usual, when I got out to the lady's house the pigeon had gone. The lady told me that she had offered the pigeon bread crumbs and split lentils, neither of which it would eat. Next day my telephone rang and it was the lady to tell me that the pigeon had come back and spent the night on the top of her porch, where it was sheltered from the rain by the overhanging eaves of the house. It remained for over a week, going away during the day and returning in the evening. What puzzled me was whether it was trying to get home each day, or whether it had found food somewhere else and simply returned to the porch-top for security at night.

Another pigeon which attracted my attention was one with a little colony of street pigeons which frequent the streets near my house. My attention was drawn to it because it was a recessive red, and even the dominant reds are few and far between among feral pigeons. There it was, complete with metal and rubber race ring, trailing along behind the other birds as they expertly searched the streets for scraps of food. It was much more nervous of passers-by than were the feral pigeons. I got quite close to it, and to my surprise it responded when I went through the motions of throwing down some food. If I had any peas or beans I am sure I could have caught it. Judging by its feet and the soiled rubber ring, it had been on the road for some time. Anyway, it only stayed a few days and then it either got on its way again or fell victim to a passing motor car or a prowling cat. I often think what a wonderful tale of adventure a pigeon which has been away for a week or two could tell on its return home! The difficulty is we see our favourites when they are in top physical condition, alert and wary of anything strange and it is hard to imagine them otherwise. Fatigue, thirst and hunger diminishes this alertness to a remarkable degree, so it is not surprising that the tired and hungry pigeon loses whatever additional urge it may have had when basketed. By the end of three days pigeons which have failed to make their loft are concerned only with self-preservation and the essentials of life—food, water and shelter. Whether the inexperienced bird finds these things is a matter of chance.

Changing Desires

Frequently, while exercising my birds in the evening, I have seen a stranger join them, and it is interesting to see how these strange birds differ in their behaviour. When I call my birds in, down comes the stranger with them, but at the last moment it breaks away, circles and is gone. Although keen enough for the company of other pigeons, it is alert and fearful enough of the strange loft. Another time the stranger actually alights, my birds trap and the stranger finds itself alone and either flies away, or flies up on to the roof top and settles there for the night. On other occasions the stranger enters with my birds. This desire for security as night approaches is very strong in pigeons: just watch how at that time they fight for their perches. Most fanciers will have experienced the prisoner which escapes, or the bird which, when you try to break it, clears off right away, neither of them even circling your loft. Yet, as night approaches the apparently lost bird is seen circling in the neighbourhood. Your pigeons are put out and joining them the lost bird then enters the loft. There is no doubt that changes have taken place in the motivating desires of these birds since they originally flew off. There is no disputing the fact that they are now hungry, thirsty and desirous of a safe perch for the night and react accordingly. We can all at times act very illogically from lack of thought. All of us, I suppose, have at one time or another had his pigeons out exercising or just picking around the loft. Suddenly we decided it was time to have them in, but after all our calling and coaxing one or two stayed out. We got annoyed and chased them, and instead of landing on the loft they took to a roof top. So we chased them from there and put out one or two other birds in the hope that they would bring the erring birds down and in. Now, in place of one or two birds out we had perhaps half a dozen flying around, and instead of those we put out influencing the others the opposite occured; they, too, became infected with whatever upset the original culprits. It is obvious that in the first place the birds which refused to come in were not hungry or at least not hungry enough to pay any attention to our calling them. By chasing them we made them excited and then to make matters worse we put out birds who had just been fed and were not used to this treatment and they became excited too. All we could do was to wait until their alarm abated and with the approach of night hope for the best. Now I always refuse to upset the others over one obstinate pigeon; a night out without food will not do it any harm, in fact the experience might be useful to it.

Environment and Instincts

Environment modifies or intensifies the inherent instincts, and the differences we see in the behaviour of various fanciers' pigeons and in racing pigeons and their feral cousins are examples of what environment can do. What must not be overlooked, however, is that while certain inherent instincts may have appeared to have changed under the influence of environment, if they are given the appropriate stimulus pigeons are liable to revert to their original form. Thus, changes in environment are liable to bring changes in behaviour. Many years ago I bought a pair of pigeons and when I put them into the loft out of the basket they behaved like creatures possessed, flying and battering themselves all over the place. They would not come near the others when feeding and so I kept them in a nest box with food and water. Even after several months I had no control over these pigeons and very foolishly tried to break them, with the result that they disappeared. Later, I learned that the fancier I bought them from gave his pigeons a free entry and exit. He never trained his youngsters until they were yearlings, when they got a few training tosses at the end of the season. When basketing or trapping his pigeons he pulled down blinds and the birds were caught in the semi-darkness. Pigeons from this loft were a perfect menace when brought into the average loft, grunting alarm signals and upsetting the other birds. More recently I received two similar pigeons, but this time I changed my tactics. To my knowledge they never picked a single pea for three days and when they did start to come down to be fed they left it so late that they got very little to eat. I can assure you that by the end of two weeks they were eating out of my hand, but I really had no control over them unless they were hungry, when they readily recognized the call to come and be fed. So, one dull evening I let them out with the others and when called in they came as if they had been doing it all their life. I took care only to let them out in the evenings when they were hungry. Closely related pigeons, no matter how much they may have resembled each other originally, will come to differ when exposed to different environments. In this connection I recollect Jack Clark saying to me, 'It is a funny thing but any youngsters I get from you are easy to recognize as of my family of pigeons. Yet I regularly get birds sent to me by other people bred from birds of my own family and only a generation removed but they do not resemble mine in the least.' It is true that Jack's loft and mine were similar in many respects and our method of training and racing varied very little.

Feeding and Thirst

What effect the environment of the basket has on the original feelings, emotions and their resultant urges we can only guess. We can be sure that what the individual pigeon experiences in the basket will influence it in one way or another. It is reasonable to assume that those pigeons which for some reason are bad travellers will go under early on or make a poor shape at racing. One of the greatest mistakes a fancier can make is to see that his birds have a feed before being basketed for a training toss or race. Pigeons require water to digest their food; in other words food in the crop induces thirst. If my experience and observations are any criterion, returning racers are keener to have a drink of water than to have food and to me the inference is obvious. I do not think that enough attention has been paid to the conditions and effects of travelling on pigeons. The question of ventilation and lack of real rest have been particularly neglected. In the armed forces practically every form of transport was used for pigeons. I found that to obtain reasonable results a quiet rest in the basket before liberation was absolutely essential. I vividly recall my amazement when some pigeons which had arrived by aeroplane were taken out on to the tarmac and liberated. They simply walked out of the basket and started preening their feathers and when they were chased into the air all alighted on the roof of a hangar, staying there for some time before taking to the wing and making off.

A Good Start

The late Arthur Stow of Barrowford who, in his day, had few equals and was so keenly observant and of such wide experience that it was a treat to me as a young fancier to listen to him, was emphatic that there was nothing to be gained by liberating pigeons before the sun had risen, and stressed the importance of a good start. He declared that a convoyer need never be in any doubt as to when his birds were ready to be liberated; their behaviour in the baskets would tell him all too plainly of their anxiety to be off. On the subject of feeding, he told me that pigeons fed the night before did not require anything on the morning of liberation and that even if food was given to them very few bothered to eat it. I do not think that there can be any doubt that many pigeons are out of the race before the strings are cut merely because of the effects of their journey. My old friend Charlie McCann of Limekilnburn, on his return from a trip to France with the birds of the Lanarkshire Social Circle, told me that he had changed his ideas about preparing birds for the journey to France. He said they must be accustomed to roughing it in the basket and facetiously declared that he was going to suspend a basket in an outhouse, place the pigeons in it, and give it a swing every time he passed and periodically throw a bucket of water over it! He declared that after such a long journey the pigeons required at least twelve or more hours quiet rest at the race point. Many of the unsatisfactory races we experience are the direct result of the pigeons arriving at the race point late and then got out of the vans in a hurry and immediately liberated. One difficulty is that convoyers are tied down by regulations passed at annual general meetings. In my view they should have a freer hand and if they cannot be trusted then they should not be on the job. Despite the vicissitudes of the journey the bulk of the birds do retain the urge to return. However, very often the stress and strain of long hours on the wing weakens them to the point of exhaustion so that their only desire is to rest. Irrespective of the opinions held on the many aspects of pigeon racing I think that it will be agreed that abundant good health, a tip-top physical condition and one of the favourable nesting conditions are essential to a pigeon performing well in any race, and that these are most likely to be acquired in a constant and suitable environment.

ESSENTIALS

REGULARITY	AIR	Good ventilation and freedom from dampness
	WATER	Fresh and in clean vessels
	FOOD	Sound food, rationally fed
	CONTROL	Exercise, trapping, feeding and basket work at the right time

OBSERVATION

3 The Art of Feeding

The variety of things pigeons will eat and apparently thrive on is surprising. However, there is a difference between the food necessary for survival and the food necessary to produce the high state of physical fitness demanded by our sport. Racing pigeons require a balanced diet, though it need not be confined too narrowly to proteins and carbohydrates. Suitable combinations of these in the form of mixtures can be obtained from any grain store, or the fancier can mix his own selected grains to suit his fancy. A great deal has been written on this subject, and *The Thoroughbred Racing Pigeon*, by John Kilpatrick, deals with it comprehensively.

Summarizing his views, he says, 'I, like many other fanciers, once thought that the richer the food the better the performance of the birds, but experience has compelled me to make drastic changes in my opinion on the subject.' Later, he states that 'the most satisfactory results with both young and old during rearing were obtained with a 19 per cent protein content. A mixture containing 16 to 17 per cent protein will give very satisfactory results in racing to any distance under any circumstances. No change should be made in the feeding until the birds are heavy in the moult when the protein content can be increased to 19 per cent. For the winter, the moult completed, 13 per cent protein content is sufficient and this can be further reduced if the hens show signs of pairing among themselves.'

I know of no one who gave more time and thought to this subject than my old friend John Kilpatrick of Belfast, and many an hour I have spent with him discussing the subject of food and feeding. The art of feeding is not one easily acquired and many fanciers have spent a lifetime at the sport without apparently mastering it. The late Billy Pearson of Keighley told me that in his opinion the man who mastered the feeding of pigeons had the world at his feet. I readily admit that, like John Kilpatrick and many others, I once thought that the richer the food and the more they got of it the better the youngsters grew and performed in racing. This conception of feeding racing pigeons has ruined the chances of more good pigeons than any other single cause. No one is more painfully aware of the difficulty of trying to persuade fanciers that there are more pigeons and races lost, and unnecessary deaths among stock birds as the direct result of overfeeding, as I am. I know that the very moment one mentions any form of rationing or restriction of feeding the word starvation is uttered.

Disease

I know really very little about the diseases of pigeons, but from what I have read and the little experience I have had, the conditions which are conducive to disease are overcrowding, bad ventilation, overfeeding and dampness. First of all we must remember that racing pigeons are kept under artificial conditions in which they are never required to search for their food as nature intended and provided for. Indeed, in the majority of lofts they do not even have to wait very long for it. In addition, they are supplied at all times with food of a highly nutritious kind, a little of which goes a long way to supply their needs and compares favourably with what they would be able to find by their own resources.

Full Crop

A pigeon's crop is simply a storage place, which allows it to pick up its food where and when it can and digest it gradually later on. It is estimated that a full crop of grain takes twenty-four hours to digest. However, pigeons kept under artificial conditions in which they are fed at regular intervals have no need ever to have a full crop. To me, therefore, the art of feeding is to give pigeons, at any one time, only sufficient food to last them until the next meal is due. The important thing to remember is that hunger is the stimulus to appetite and pigeons that never, in the twenty-four hours, feel the pangs of hunger rapidly become sated with food, listless and lazy. I have suffered the disappointments of possessing a loft of fat, lazy pigeons, covered in bloom and looking the part, but which on race days arrived home after everyone else had timed in or, worse, got home days and weeks after, the very pictures of misery. Gone were the plump round bodies that I and my friends had admired so much!

Horse Beans and Maize

I remember vividly the day enlightenment dawned; I was not told since I knew too much about pigeons to be told anything. It happened in this way. As a member of the club committee, it was my turn of duty to visit the loft of the race winner and verify the winning pigeon. Now this particular member of our club was having a very successful season. On my first visit to his loft to verify a bird I asked him from where his pigeons originated and how he managed them. I was somewhat surprised and disappointed to find that his birds had been picked up singly here and there; I do not think he had bought a single one of them. As to their pedigrees he had no information. This fancier, I thought, knows very little about pigeons, especially about the finer points that I know all about. A week or two later I visited him again to verify yet another winner. I was preparing to leave when he said to me, 'Wait until I feed my birds and I will come over the road a bit with you.' He then produced a paper bag from under a box in the loft and started to throw down to his birds a mixture of horse beans and what seemed to be very large maize. When I mentioned the size of the beans and maize to him he said he got them from the pit where he worked with the ponies. At the rustle of the paper bag every bird was down on the floor scrambling to be fed; not a bird stayed on its eggs or with its youngsters, and in no time all the food was gobbled up and they were looking for more. The fancier, however, put the bag away under the box and turning to me said, 'That will do them until tomorrow.'

An Experiment

When I left him I pondered over what I had witnessed, and thought of the beans my birds refused to eat and which I swept up and gave to my father's hens. I also thought of the birds I lifted off their nest at feeding time, fearing they would go hungry before the next feed, and of the little pots of food in the nest boxes of those birds feeding young, which were scattered all over the place and trampled into excreta. I started thinking again and then experimented. Firstly I started cutting down the food ration until my birds began eating the beans in the mixture of beans, peas, tares and a little very small maize. The moment I saw a single bird start testing a bean before swallowing it I stopped feeding them. The first reaction to this treatment was that my birds became a degree tamer and obviously welcomed my arrival in the loft at feeding time. When let out for exercise they were at first a bit too keen to come back in again to be fed. However, if I kept them going for ten minutes they got into a free swing around the loft and appeared to enjoy it. When called in they trapped much better than before, but old habits die hard and there was the odd loiterer at first, but as they got nothing to eat they soon ceased to loiter. What pleased me most was that my birds obviously appreciated their food when I gave it to them. Moreover, I knew exactly how much they were getting in terms of ounces per head per day, something which I had never known before and which I have found very useful ever since. In January, once the birds were through the moult, I started gradually reducing the ration until I had to stop because some late-breds I had were beginning to show signs of distress. I had found the difference between undernourishment and overfeeding, however, and had achieved the happy medium between the two.

The Result

During the previous seasons' racing I had never been in the first ten in the club. During the season following my discovery, out of the thirteen races I won seven firsts and four seconds against strong opposition. In one race there were only four birds timed, and I was third. In the last old bird race, I timed the only bird. Even these results did not satisfy me; I had got the bit between my teeth and was determined to be the master. Some of my old birds would always alight on the house-top before coming on to the loft and one or two, when forced to exercise, alighted on a large building from which I could not chase them. I killed the latter birds out of hand. I took good care that my young birds of that season were never allowed out with the old birds to learn their bad habits. These young birds and those of the years to follow were under my control and dropped direct to the loft when called in from exercise or returning from a race. The method employed was they were never allowed out for exercise unless they were hungry. At the end of their exercise they were fed and the laggards got nothing or very little, depending on whether I was still feeding when they came in. This rule was inflexible and well worth while for what a young pigeon does when it alights from exercise it will do automatically when it returns from a race. In fact, habits formed in those early impressionable months of their lives stick to them for the rest of their days. What I now relate are only two of many similar incidents which illustrate the point.

The Laird, a very good racer over all distances, was one of the first birds I bred at Old Neilsland. My loft was in the roof of an outbuilding and for the first two years the young birds trapped through an ordinary skylight on one side of the roof. Later, a proper trap was built on the other side of the roof through which both old and young birds trapped. In a race from Dol 508 miles, when he was four years old, *The Laird* dropped at 7.40 p.m. after twelve and a half hours on the wing. I ran up the stairs into the loft waiting for him to come through the trap. After a minute or so I heard a scraping at the skylight and there was *The Laird* trying to get in. I chased him away but he only moved away on the roof. In the end I had to dismantle the wire netting covering the skylight which was now only used for ventilation, and in he came to be timed at 7.51 p.m. winning 9th L.S.C. 35th open.

The other incident occurred in a very hard race. Again at

about 7.30 p.m. I saw a bird approach and I went into the loft to be ready. After waiting a few minutes I came out to see where it was and found it sitting beside the skylight. What made these two birds make for this skylight which had not been used as a trap for three or four years and could only have been used by them at the most of six months while they were young birds. I concluded that these two birds, after their long hours on the wing, were perhaps somewhat dazed and acted automatically and their early training and habit came out on top at that critical moment. I attach great importance to the management of young pigeons during those early formative months of their life. The environment they live in, the habits formed therefrom, in fact, every experience which comes their way has a bearing on their future behaviour

Individual Feeding

I have said that the art of feeding pigeons is to give them only enough food at one meal to last until the next, but there is more to it than that. The putting down of a given amount of food at regular intervals is only a small part of the art of feeding. Young pigeons must be taught to appreciate their food and the presence of their feeder. It is essential that they should come to associate you in person with being fed. What I mean is that they may only associate the noises you make with being fed and not you personally. I make a ritual of feeding with both old and young birds, and young birds in particular. It gives you an opportunity to get to know them and they in their turn come to know and have confidence in you. First there is the rattling of the corn in the can and the whistling and calling them in from exercise. If they are being fed in the loft without having been out you should still go through this corn rattling and calling routine. It has a psychological effect; it is an appetite stimulator if you like. When I start to feed I put a little grain down and as they come to get it I push them gently away with my hand. This engenders a certain amount of excitement and you soon have them scrambling to get their share. If there are any birds outside or sitting you will be surprised how quickly they come in on hearing this scrambling noise. This excitement whets their appetites and you will be astonished at how the wildest or shyest of pigeons becomes carried away and joins in the scramble for food. Having created this excitement I feed them slowly, picking up the odd pigeon or simply pushing them aside, something to which they soon become accustomed and cease to resent. I stop feeding when I have given them about two-thirds of their feed. Most of them will then take a drink and while still looking for more food, they may have a pick at the grit or minerals. This pause in the feeding allows the initial excitement to die down and with their appetite partly appeased only those that are really hungry will come forward to eat out of your hand when you give them the balance. As they come around me, pecking at my hand, I can touch their crops and ascertain just how much each has eaten, taking care that those which have had enough get no more and giving those youngsters that may be slow eaters a chance to get their share.

This amounts to individual feeding. I can sit there in the young bird loft with a dozen youngsters picking at my hand and feed only one of them. I simply take a bean between my first finger and thumb and hold it out for the youngsters to pick from between my fingers. However, I only release it when the bird I want to have it takes it. This may seem a lot of trouble, but it is well worth while. It has been carefully reasoned out and proved in practice. What happens is that every bird has enough and, most important, none of them gets too much. Young pigeons treated in this way right from the start become reasonably tame and without being a nuisance they are handled easily and trap at once on being called. Any birds which do not achieve this standard are given no consideration. There was a time when I would not give a late trapper a single pea, no matter how many meals it missed. I have found that these pigeons become so hungry that they appear not to know what is the matter with them. Now the late trappers get a few beans, which stimulate their appetites and so they will trap after the next exercise. I detest wild or excessively timid pigeons and go out of my way to overcome their fears. With this method of feeding, the wild or timid pigeons do not get much to eat at first, and it is to them you give a little extra attention in the latter stages of feeding. Soon they will be picking at your hand for more, a sure sign that they are gradually overcoming their fear.

Control

After the young pigeons have been fed and taken to their perches and are settled for the night, I get as close to them as I can, and stand perfectly still. The best time to start this is when the light is failing, for they are less likely to attempt to elude you by flying from their perches. Once they get used to me I touch and stroke them. Some will peck at your hand or strike you with their wing, which is all to the good. In no time, however, you can pick any one of them off its perch. Do not expect them to sit and allow you to do this when they are waiting to go out or to be fed. There is a right time for doing everything and, at first, only handle them at night when they have just been fed. When you pick a pigeon from its perch to examine it do not throw it down on the floor or, as I see some people do, let it fly out of your hand. Put it back on the perch you took it from. There are several reasons for this. First, if you throw it down on to the floor other birds will also fly down. If you allow it to fly out of your hand it is unlikely to find its perch and you have it scrambling all over the place disturbing the others. Often this fluttering of wings scares some of the others, and soon you have a few birds flying around your ears. I am not suggesting that you should be habitually catching and handling your pigeons. What you should do is to get these young birds into the habit of controlling both their natural fears and the impulse to escape from you. My pigeons will rarely let me touch them unless cornered or when they are nicely settled on their perches, but even if they stay out of my reach, they do not stampede by flying away. They simply stay out of reach by walking away, without any flurry or panic. I have gone to some pains to get my birds to use their legs within the loft, instead of their wings. The doors between compartments are kept shut and the birds have to walk through small apertures at the foot of the doors or compartments which only allow one bird through at a time. Roomy lofts or compartments in which the birds can fly around have a tendency to make them wild—at least, they use their wings and are difficult to catch.

Full Crops are a Handicap

I never attempt to catch a pigeon unless I am sure I will not miss it. Each time a pigeon eludes you it is encouraged to try to dodge you again. This is particularly true of young birds. When it comes to the basketing of young birds, take the basket into the compartment, place the basket on the floor and put a feeding trough beside it. I put some millet or other small seed in the trough and as the birds scramble for it I pick them up and put them into the basket. I take care that I pick up the most wild and shy birds first. I use millet or some other small seed because I do not want these birds to have any solid food in their crops and because these small seeds take longer to pick up. Pigeons sent training or racing with food in their crops are unnecessarily handicapped with the additional weight, plus the fact that they will become thirsty. During flight, digestion is suspended, as witness the watery droppings of the returned racer. Pigeons drink a lot of water in the digestion of their food. My loftman at Old Neilsland never ceased to be amazed at the quantity of excreta he scraped up and carried out of the loft, declaring it was always more than they ever got to eat. What he overlooked was the water they drank. One season, while helping a friend to basket his birds for a training toss, I told him that there were four birds in the basket that I would not send because they had far too much food in their crops. He replied that they were only going ten miles. But that night there were three missing and they were three of the four with a lot of food in their crops. They all turned up next morning and still had food in their crops. If you doubt this, just let one of your birds get a crop full of food and then let it out for exercise with the others that have empty crops. You will see it trail behind and watch it try to turn. These may seem trifling details, but successful pigeon racing is made up of the observance of detail, none of which is all-important, but if neglected can throw your programme out completely.

Feeding Young

For many years, when feeding young birds in the nest, I followed the popular practice of placing a dish of food in the nest box of parents which were feeding youngsters. The general idea is that it ensures that the parent birds always have enough to feed their young. Moreover, the young, seeing the parents picking, learn to pick for themselves. I discontinued this practice for several reasons. The first was that I did not find that parents so treated reared better youngsters. Second, it led to a lot of waste through the food being scattered and fouled. Third, the other pairs, seeing food in these nest boxes, would enter them and this led to a great deal of unnecessary fighting. Fourth, I noticed that when feeding parents with food always in front of them they became sated and lost form, and having full crops they seemed to feel that their young were in the same condition and neglected them. Finally, parent birds with full crops of partially swollen grain appeared to have difficulty in regirgitating food and their young had equal difficulty in swallowing it. The time for the parents to feed their young is when the grain is freshly eaten and they have just taken a drink of water. Youngsters easily swallow the unswollen grain and the freshly taken water helps the process. The place for the grain to swell is in the youngsters', not the parents', crops. A youngster getting a half crop full of freshly eaten grain with lots of water will, in a comparatively short time, have a crop so packed that another bean could not enter. Moreover, it is much easier on the parents in every respect.

Olympic Method

My method is to feed the parents in the normal way along with the other birds that are sitting on eggs. After they have been fed and have taken a drink their first concern is to feed their young. I watch them do this and now and then have a quiet look to see what kind of job they are making of it. The parents with big youngsters will have emptied their crops completely; those with smaller youngsters to a lesser degree. The important thing is that the young have been fed with freshly-eaten grain and water which, in combination with the warmth of the young birds' bodies, will have swollen surprisingly in a matter of an hour or so. After the parent birds have fed their young I allow a period to elapse, busying myself with the odd jobs about the loft—taking notes of eggs laid, ringing youngsters or just quietly watching the birds. I now take my feeding dish and give each of these pairs with young some extra food, basing the quantity on the size of the youngsters and taking care that it is all picked up. Those with youngsters under a week old get no extra food. Pigeons treated this way rear healthy young, and remain alert and active themselves. I have no time for a pigeon that cannot rear a pair of lusty youngsters and take the whole proceedings in its stride. My method ensures that you will not have pigeons suffering from sour crop or other digestive disorders, as their crops are emptied regularly and are never distended for any great length of time. I do not think there is anything more distressing than to see a pigeon with a drooping and distended crop, the feathers of its breast soiled and wet with the struggle it has had trying to get the swollen grain over the throats of its young. It is rather amusing when you have a loft of pigeons which have become accustomed to my method of feeding. Those birds which are sitting on eggs will, of course, have been fed in their nest boxes when they were feeding young. Naturally enough when I enter the loft to give those feeding young their extra ration, practically every bird makes a bee-line for its nest box. To keep those that are not feeding young busy I put a little millet in their nest boxes, and every bird is happy. Again, as already described in the feeding of young birds, the feeding is almost individual and the birds are never in any doubt about who feeds them. The extra time spent feeding your birds is time well spent. It gives you the opportunity to observe many little things which might otherwise pass unnoticed. The bird which seems more than ordinarily hungry, or the bird which is disinclined to eat, both require investigation.

When Not to Feed

All the fine points of feeding would fill a book. There is always some situation cropping up which requires some thought. There are so many likenesses and differences. The bird with the full crop does not always mean it has had too much to eat. It may be suffering from a stoppage in the proventriculus, an impacted crop or gizzard, or other digestive disorders. Pigeons which have been out a night or two should be fed very carefully, since they are inclined to gorge themselves and upset their digestive system. On the other hand, pigeons that have lost their appetite are always suspect: watch carefully the birds which keep picking around the loft floor as if looking for something. It is not wise to feed pigeons close to darkness because after they are fed they take a drink of water and an hour or so later they will require another drink in order to digest what they have eaten. If you feed so late that they cannot get this second drink then their digestion is impaired. If the night is not a very dark one or if there is a moon, some birds will come down from their perches in order to drink. This will bring the others down and you will find that they will have spent the night on the floor. Some, perhaps, may have fluttered and scrambled on to the perches. This sort of thing disturbs all and sundry. If you have to feed late at night and have electric light in your loft, switch it on for ten or twenty minutes an hour or so after feeding; this allows them to get a drink and return to their perches. Regularity in feeding is a virtue but never fear that missing a feed will do your pigeons any real harm.

Post-mortem Results

Every now and then I have pigeons brought to me that have died or are obviously seriously ill, to see if I can diagnose the cause of death or their complaint. I will only deal with a few cases in which the primary cause of trouble was connected with feeding. Returning home one evening my wife told me there was a pigeon in the kitchen. I asked who had brought it and gasped with astonishment when she said the postman had. It turned out, however, that it was a pigeon the postman had picked up in the street. I examined the bird and found that although it was emaciated its crop was packed full. The contents of the crop felt like a doughy mess, which I tried in vain to break up. I gave the bird some milk and cod liver oil and put it in a basket with water to drink. The next day there was no improvement and it had passed no solid excreta. The following night it was obviously dying; its eyes had lost their life and colour. I killed the bird and found that the contents of the crop appeared to be a solid mass of barley or some other cereal. This had become glutinous, refusing to break up and pass through the proventriculus to the gizzard which was entirely empty, as were the intestines. Thus, here was a pigeon which had died of starvation in spite of its crop full of food. How much the lack of grit had to do with the condition I do not know. In addition to the grinding of the food in the gizzard, grit has a secondary function. It helps to keep the food particles apart in order that the digestive juices can get at them and prevents them becoming impacted, as had happened in the case of this pigeon. I have at different times had a number of dead youngsters brought to me. In every case they were healthy well-reared youngsters who had been found dead in their nest boxes, with a crop full of peas. It was the practice of all their owners to place a pot full of peas in the nest box of all feeding pairs. The victims had all learned to eat for themselves and did not trouble their parents to be fed. They died for the want of water. What surprised me most was that they died very quickly, being in excellent body condition and not emaciated in the least. The other cases are comparatively common and most fanciers will have had the experience of finding a bird dead which to all outward appearance was the picture of health, covered in bloom and which had shown no signs of being off-colour. The cause of death in these cases is internal haemorrhage, usually from rupture of the spleen; and in some cases the liver and kidneys are also involved, being the result of the progressive fatty degeneration of these organs usually caused by overfeeding and lack of sufficient exercise.

How Much is Enough

The rational feeding of racing pigeons is an art. I would be delighted to be able to set down the exact amounts to be fed per bird per day for every occasion. But, like the protein, carbohydrate and mineral contents of the food we give them to eat, their needs vary from day to day with the work, weather and other circumstances. Like ourselves, one day pigeons will eat more than another day. On top of all this, what would satisfy the pigeons in one fancier's loft would appear to be a starvation diet to the birds in another's loft. The reason for this is that pigeons, within certain limits, eat as they have become accustomed to eat from their earliest days. I made some tests of this and had some pigeons killed and examined by a veterinary surgeon and it was found that the birds that had been given as much as they could eat developed larger gizzards than those kept to a fixed ration per day.

Racing pigeons can be kept in top physical condition on an average of one ounce of food per bird per day. Youngsters of one week old or more must come into the reckoning. During the months of January or February the allowance per bird per day should be at the lowest. It should be raised when the pigeons start feeding their young and reduced slightly as serious racing comes along and most of the birds are sitting on eggs. I cannot recall ever possessing a pigeon which was a big eater and any good at racing. It is athletes we want, built and tempered for speed and endurance. Pigeons rationally fed and exercised will thrive on around one ounce per head per day and they will be found to be more active and vigorous at all times.

Thoughtlessness Changes Environment

After I had spent a considerable time during the racing season teaching a young fancier the art of feeding, with gratifying results in racing, I felt confident that he could be left to his own resources without slipping into any of his old habits. After his birds had moulted he asked me round to discuss with him his matings for the next season. On entering his loft I got a surprise. I asked him what he had been doing with his birds. He asked me what was wrong. I answered, 'Your birds are far too wild and I know the cause.' During the very cold weather I could visualize him rushing down to his loft, giving it a hurried scrape out, refilling the water fountains, throwing down their food and clearing off as soon as possible out of the cold. His birds had ceased to welcome him as the provider of their food. Instead he had become the whirlwind that entered the loft, chased them around as he scraped it out and filled the fountains, and only then did they receive their food and have the peace to eat it in. I told him to cut down their ration, bringing it well under an ounce per bird per day and to spend some time feeding them by hand. I have the notion that he thought I was being a bit drastic considering the cold January weather. The sequel was that meeting him some ten days later he informed me that his birds for several days now had not been eating all their ration. Pigeons never take kindly to sudden changes of any kind and take some time to adjust themselves and this is most noticeable with any change of diet. I say this because I fear that so much that is written consists of opinions culled from what the writers themselves have read or been told, as opposed to findings garnered from their own personal experience and investigations. Otherwise much of the cant and pseudo-dicta so often repeated would have long since disappeared from present-day literature. I was amazed to hear a prominent fancier and writer express the opinion that all that could be written about pigeons had already been written. Just think again of the many things we would like to know but which so far as evaded us, even about the apparently simple subject of feeding.

Food Comparisons

Compared with the amount of scientific research that has been done in connection with the feeding of cattle and poultry, very little has been done concerning racing pigeons, and fanciers have been left to their own resources to find by trial and error what appears to be best. Feed them on the best that can be bought and judge the best solely by the price you are asked to pay for it. On top of this we have inherited from the past the opinion that the richer the food the better it is for pigeons. I once thought it strange that although I was feeding my birds on a mixture of tic beans, maple peas and tares with a protein content of 22 per cent for which I paid £1.30 per cwt., I was being beaten every week by a fancier who fed horse beans and large maize with protein content of around 16 per cent at a cost of about 40p per cwt. At about the same time I recall the furore created when it was discovered that a fancier who was hitting the high spots was feeding a white pea imported from China which he was buying for 30p per cwt. Immediately there was a run on the white China pea until a lot arrived riddled with weevils.

The answer is, of course, that pigeons will do well on a great variety of grains provided they are well matured and sound, and contain a nice balance of protein, carbohydrates, fat and ash (mineral). It appears that a pigeon's powers of digestion of protein and starch are extremely good, since they are more efficient than poultry. However, they are less efficient in the digestion of the non-starch carbohydrates and fibre which passes out undigested. This is confirmed by the fact that when fed on beans and/or peas with a fibre content of 7.1 and 5.4, respectively, their droppings are firm. But when fed with maize and/or wheat with a fibre content of 2.2 and 1.9, respectively, their droppings tend to be loose. This is because the undigested solids reach the cloaca in a comparatively firm condition and when the fluid waste in the form of urine reaches the cloaca the solids there absorb it completely, producing firm droppings. If the solids in the cloaca (fibre and other waste matter) are unable to absorb the urine completely then we get loose, watery droppings. Failure to understand this fact leads fanciers into thinking that maize and wheat scours their birds. Another example of this is the watery droppings of the newly returned racer which has been hours without food.

Vitamins and Cod Liver Oil

Pigeons require more minerals than poultry. Fortunately, there are sufficient of the so-called trace elements in the grain we feed them on to meet their requirements but there are other minerals which require to be provided as supplements to their normal food. These are usually provided in the form of grit consisting of oyster shell, flint and limestone, not forgetting salt. Bone meal is also recommended as the best source of calcium phosphate. Vitamins are adequately provided for in the most common grains fed to pigeons. A useful supplement, particularly in the autumn and winter, is cod-liver oil (see Appendix C).

Feeding Ritual

Successful pigeon racing is made up of the observance of many little factual details and the recognition of the likenesses and differences. In this respect it is not so much what is done but how it is done. The putting down of the right amount of food at regular intervals is a step in the right direction but it is not enough. I make a ritual of feeding and have fully explained fully my reasons for doing so. The situation is a psychological one bringing into play a number of rather complex factors. The chief thing is that it works in keeping racing pigeons alert, active and healthy at all times. The interesting thing is that pigeons fed in the manner I have described will actually eat more than if they were hopper fed with food always before them. The reason is that they have a keener appetite and appreciation of their food when offered to them and, equally important, is the person who feeds them. How often have you heard a fancier say that he had stopped racing his birds intended for the long races and that all he needed to do was to get some body on them. I can well imagine that his intention is to do this by feeding and by coaxing his birds to eat make them lose their appetite and form. I am absolutely convinced that the chances of many a good pigeon is ruined at this time. The plan is to sit tight. The rest from racing and proper care ensures that they retain their appetite. That is all that is required, and if they are the right kind of bird they will have all the body reserves necessary for the job in hand.

Body and Weight

All the really good racers I have handled were buoyant pigeons, handling bigger and lighter than they looked on the perch. This fetish of getting body on them is one of the apparently undying myths.

I often have heard the unsuccessful exhibitor at a show ask how the owners of the winners managed to get a full rounded body and such wonderful bloom on their pigeons. The answer is that these pigeons are bred to have such bodies and bloom, and within certain limits no amount of feeding or other treatment will put body and bloom on pigeons which do not possess them by inheritance. These lovely-bodied show racers tend to put on weight which is another matter and it is not unusual for these birds to get practically nothing but water to drink for twenty-four hours before being exhibited, or a dose of Epsom salts to empty their intestines in an attempt to give them some semblance of buoyancy by a reduction of their weight. The newcomer, having listened to all the talk about body, will handle a few pigeons whose bodies his friends have admired and praised. He will no doubt, at some later date, be surprised and perplexed on handling a pigeon with an excellent racing record which by comparison with the others had no body to speak of. Thus, he will learn that all pigeons with good bodies are not good racers and vice versa.

We are all aware of the relationship between body-weight and health. Emaciation is immediately recognized as an indication that something is amiss, and we take steps to find its cause and, if possible, to put it right. Yet overweight is just as clear an indication that all is not well. However, it is very insidious. Its victims look all right. You may recall the sudden deaths, the birds you thought were fit that returned late or were never seen again and the escaped prisoner that is never seen again. These, in many cases, are its victims. For one or two racing seasons I weighed my pigeons before and after the races, noting the variations from race to race; but I abandoned regular weighing of my birds for two reasons. First, because the scales I had were not accurate enough for the purpose and, second, I had not the time to do the job properly. I found that there is a very fine relationship between body-weight and top form; so fine that it cannot be ascertained by handling, but only by carefully weighing the pigeons first thing in the morning when they have empty crops. With these pigeons which have been sitting on eggs or small youngsters, one must wait until they have evacuated the droppings accumulated while sitting. You will be surprised at the weight of the excreta of a hen that has been sitting all night. Also, a suitable balance weighing accurately in drams can be a quite expensive item.

Weight and Form

I found that pigeons which lost weight very readily were never much good. Fit pigeons on the other hand varied very little in their weight before and after being on the wing for up to seven or eight hours. However, as I have already said, my scales were not accurate enough under an ounce, and there might have been a difference at that level. Pigeons which were being raced week after week but still kept their weight or made a slight gain were usually good propositions to pool. I also found that a good stiff race involving twelve to fourteen hours on the wing, or even a night out, seemed to bring certain pigeons into better form than they had previously shown. I have noted this over and over again and had it confirmed by others. My conclusions are that it required this stiff test to bring the birds concerned down to their best racing weight and physical condition.

Gain and Loss in Weight

To my way of thinking body and weight, while associated, are two different things. Each individual pigeon by its inheritance conforms to a given form, shape, structure and body—the flesh or muscle covering its frame. These factors vary little if any during the lifetime of each pigeon under normal circumstances. Weight, on the other hand, can and does fluctuate perceptibly. During the autumn and winter months pigeons put on weight, and the great mistake is to confuse this excess of fat as being an increase of body or muscle. Rationally fed, and given work in the form of exercise, training and racing, the excess fat disappears and the physical condition of the pigeon improves. Some pigeons require to be worked hard to keep them in good physical condition while others do not seem to be affected to the same extent. Irrespective of what sort they are, this feeding them up to put so-called body on them is a myth. Often I have witnessed both my own and others' birds returning worn out and emaciated but after a rest and a return to normal body-weight able to race again and win handsomely. I like a bird to have a good body but not weight—weight is a handicap.

Regularity

Regularity in feeding is essential. I have witnessed more upsets from sudden changes of feeding habits than any other cause. Let me give a few examples. A fancier who has been feeding his youngsters, both in the morning and in the evening after exercise, decides one Saturday afternoon to let his youngsters out as he would like to go somewhere in the evening. The result is that he is still waiting for some of them to return when evening comes. Next morning he lets them out as usual and is surprised to find that again a number clear off and he has to wait for hours to get them in. The reason for this behaviour is that on the Saturday when let out they were not hungry and went roaming. When the fancier coaxed the latecomers in on Saturday, those that were already in got too much to eat and still had food in their crops when let out on Sunday morning and so the performance was repeated. Another example of the same thing resulted in a flyaway and the loss of a dozen youngsters.

Feeding Once Per Day

Personally I have given up feeding my youngsters in the morning. They get one feed per day and that is in the evening after exercise. This arrangement allows me to at the shortest notice to take every available opportunity of giving them a training toss without food in their crops. In time they come to learn that arriving home means being fed and they are ready for it. Exercising and feeding only at night reduce the chances of them straying, as they have two very strong incentives to enter the loft; first, to be fed and second, their desire for their perch and security as night approaches. Youngsters treated in this way and used to a long fast are less likely to be upset by lack of food when later on they may have to spend long hours on the wing, returning from a bad training toss or race. It is all part of their training to become racers: they must have what it takes or they will not be much use to you.

I have explained very fully my ritual for feeding: now and then even the best of us can be caught out by a lack of thought, or the unexpected. One evening my youngsters had just alighted and four of them had entered the loft. As I threw some food down there was a loud bang and the youngsters that were still outside took to the wing again. After a bit of calling I eventually got them all in. As is my practice, I started handling some of them when they had settled on their perches and to my annoyance I found four of their crops packed full. In fact, these youngsters still had food in their crops the next afternoon. The moral of this is to get the majority of your youngsters in before putting down any appreciable quantity of food. A little millet or other small seed will keep the first in busy until the majority are safely inside the loft, when you can start giving them their regular feed. If there are one or two still outside they will have to take their chance. The main thing is that no pigeon gets too much; the greater the number the more even the share. Writing this recalls an amusing incident.

Response to a Challenge

In the early days of World War II we were using pigeons of several local fanciers and I called on one of them on a Saturday, in time to see his seventh or eighth bird arrive from a race. After it had entered he asked me whether I could pick out his first arrival. This was a challenge I took on. I handled the seven or eight youngsters and without any difficulty placed the first four correctly and then with some difficulty I placed the others. He was amazed, but I was surprised that he did not ask me how I did it. I had seen him get one pigeon in by throwing down food in which all the earlier arrivals shared and it was simple logic to assume that the first bird home would have more food in its crop than any of the others. Thus, by judging from food in their crops I placed them in their order of arrival. My own practice is to give each bird as it arrives the amount of food I want it to have in the trapping compartment before letting it in beside the other birds.

I remember another story concerning this same fancier. One

Friday afternoon a call came for some pigeons to be sent out to an Army unit. As he was the nearest, I thought I might catch him before he set off to the marking station with his birds. When I arrived he actually had them basketed. After he had given me the old pigeons I requested, he insisted that I handle the young birds he was sending to the race. So far, he had been racing very well indeed, and as he handed me each pigeon I was told how it had performed and how he had it pooled for the race the next day. Among the last birds I handled was a Blue Chequer hen and as he had not pooled it I asked him if he had any objections to me having a flutter by pooling it. He readily agreed. During the following few days I was away in another district but on my return the driver who regularly collected the pigeons for the Army told me this fancier wanted to see me. When I arrived at his house his face was all smiles and after greeting me, he said, 'Here are your winnings.' I do not recall whether the pigeon I had picked had won the race but it certainly was first in the pools. He asked me how I came to select this particular bird in preference to others with much better records in previous races. I told him I chose it simply because I thought, all other things being equal, it stood a better chance the next day than any of the others. My reason for so thinking was that it was a second-round youngster and in much better feather than the first-round youngsters that had been winning in the previous races. These were now very ragged around the head and neck, their secondary cover feathers and tail feathers were out and the others were only partly grown. The forecast for the next day was a west wind with frequent showers. Acutely conscious at the time of the effects of over-full crops, I had noted that for some reason or another his fancied birds were simply packed with food, while this Blue Chequer hen was just right.

When to Stop Feeding

While I am all for racing pigeons being reasonably comfortable while in the basket (where water is essential), pigeons basketed on a Friday evening for liberation the next day should not be fed. When in the basket longer than this they should only be fed in the late afternoon or early evening and certainly not on the morning of liberation. This, I know, will in the opinion of some amount to sacrilege, but I would much rather have my pigeons without food than overcrowded in the race basket for the period of waiting. The essentials of life are air, water and food in that order and the last without the first two is worthless. Many successful fanciers in all honesty will tell you that they give their pigeons as much as they can eat. But ask them how much they feed in terms of weight per bird per day and they cannot tell you. Being deeply interested in this question of feeding I take special note as to how successful fanciers feed their pigeons. Irrespective of their method of feeding the striking thing they all have in common is that they knew when to stop feeding, and therein lies the art of feeding. Many know this and keep it a secret. Others have been told, but are possessed of that great fear, which is the cause of most of their disappointments. Georges Fabry says, 'It is a great mistake to overfeed your pigeons thinking it keeps them healthy. It is much better that they have a good appetite and are always looking round for more.'

The Importance of Regularity and Uniformity

If you think again before throwing down that extra handful of feeding and put it back in the corn-bin, it will not only save you money on feeding but will win prizes for you out of turn. Cultivate regularity in feeding. By this I do not mean only in your time of feeding but also in the amount given. Feed once or twice per day, giving at each meal only sufficient to last until the next is due. Normally the morning feed will be less than the evening feed because the hours between are shorter. Unless you know exactly in terms of weight per bird per day what you are giving your birds, regularity cannot be achieved. The hermatically sealed tins in which a ½ lb. of ground coffee is marketed holds 1 lb. of a mixture of beans and peas and is enough per day for sixteen pigeons. Similarly, the normal sized teacup holds five ounces or enough for five pigeons. The one-pint household cooking measure holds a pound of beans and peas. However, be careful, for some of these measures are graduated in fluid ounces of which there are twenty to the fluid pint.

Throughout this chapter I have written in terms of feeding on the natural hard grains normally used to feed pigeons. The poultry industry has spent a lot of time and money in research and has produced pellets to meet every stage in the life of poultry, the end products being the laying of eggs or to fatten for eating. With the feeding of racing pigeons the aim is to get and keep athletes fit and healthy. I am at the moment not convinced regarding pellets for pigeons, not having seen any minutiae of their contents or of the tests made as to their suitability. This I do know: that birds fed on them drink more water than otherwise, suggesting excessive thirst—a serious handicap to pigeons when racing. Reading between the lines of the adverts by the big firms specializing in pigeon feeding, I wonder on what sound scientific basis they and the other purveyors of supplements, tonics and stimulators base their wonderful secret formulae. The trace elements, as their name implies only require to be present in infinitely small quantities and an excess can be as harmful as a deficiency. You do not require to keep a bonfire burning to light your cigarette or pipe, when the small flame from a match or lighter serves the purpose.

I trust the reader will try to master this chapter as a whole and not take the odd paragraph out of its context. It would be asking for trouble suddenly to cut the feeding to a ½ once of wheat per bird per day if previously your birds had been eating their fill daily. If and when you start to rationalize your feeding, remember it will take two or three weeks before your birds settle down to their new regime. Old birds do not like changes no matter what; they must be made gradually: young birds are more adaptable.

A carefully-thought-out and constant routine which works in with your hours of work and leisure pays dividends.

Should you find, after cutting down on the amount of food being fed, that your birds when put out for exercise keep trying to come back to the loft, don't worry, it simply means that they have not yet become used to the change.

4 Training

PART ONE: YOUNG BIRDS

In the last chapter the art of feeding was fully discussed. We saw that young birds have to be taught to eat and fend for themselves, to appreciate their food and become accustomed to their owner and to realize that in the loft is to be found everything they want: food, water, security and the company of other pigeons. Young pigeons let out for exercise fly for the sheer joy of it, in the same way that lambs skip and kittens play. Their feral cousins, however, from an early age have all their time taken up searching for food. They do not circle and fly aimlessly around, for life is much too serious a business for idle flying if they wish to survive. Young racing pigeons have no worries on the score of finding food, so from the lack of somewhere to go they circle their loft and if carefully trained will come in when called to do so and be fed. If they are allowed a free exit and entry, they will fly around the loft and only enter it when they feel hungry, thirsty or consider it is time they found their perch for the night.

Teaching Pigeons to Drink in a Basket

Before young pigeons are taken for their first training toss it is a good idea to give them some experience of being in a basket. There are many who go to great pains to get their youngsters accustomed to eating and drinking in the basket. Some in fact wean their youngsters in a basket. Personally, I have no wish for my young birds to like being in a basket. In my opinion, there is only one place a pigeon should want to be and that is in the loft. I do, however, attach some importance to teaching my young birds to drink in the basket. The method

I adopt is simple but effective. One evening when they are out having their exercise and when I think they should have their first training toss, I remove the drinking fountains from the loft. At the termination of their exercise I feed them in the usual way. I then leave them for a time knowing that they will become desperately keen for a drink of water. I then basket them, allowing them time to settle for a little before placing dry water troughs on the basket. I pour water into these with much splashing and noise to attract their attention. It is surprising how quickly each bird sticks its head out and drinks. Indeed, you might think it had been doing it daily. Like most things in the proper management of pigeons, this is all a matter of timing. I make sure they are thirsty before I introduce them to it. I make no attempt to teach them to eat in the basket. Unfortunately, current ideas on how pigeons should be treated when sent racing mean they will learn soon enough. In the meantime, I want them to associate being fed with the loft and certainly not with the basket.

Exercise Basket to Loft Only

I have been fortunate enough to have had the opportunity of training pigeons under a great variety of circumstances and with different objects in view as compared with the orthodox methods practised for racing purposes. As a consequence I have accumulated a great deal of knowledge concerning the effects of environment and training on what we have come to think of as the normal habits of racing pigeons. Many of these habits which we look upon as normal are in fact purely environmental. There is no doubt whatsoever that the circling of their loft at exercise and part of circling on being liberated are habits brought about by the environment in which they live. Change the environment and the circling around the loft will cease, and alter your methods of training and the circling on liberation will practically disappear. During the last war pigeons were once required at short notice for an operation to take place in about ten days' time. The only birds available were two lofts just freshly stocked with youngsters, six to eight weeks old. They were just broken to the loft and were beginning to take to the wing. Training was started immediately and the first toss was made within sight of the loft. For the first day or two they were tossed at about half a mile from the loft three to four times a day. They were then taken twice a day at gradually increasing distances in the direction in which they were expected to operate. Within ten days they were flying from thirty miles in pairs and trios, and losses were practically nil. These youngsters were never exercised around the loft as they got enough exercise in their actual work. When liberated from the basket they went off with the minimum of circling and they rarely rose to any height; to see them arrive and trap was a treat to watch. One surprising thing was that later on when these pigeons were in company with pigeons trained and exercised in the usual manner, these youngsters would separate out of the flock and set off for the loft on their own, leaving the others circling. They usually made better times than the other pigeons and were more reliable when tossed singly. Few fanciers have the time or the facilities to carry out such a programme of training, but most could, if they cared, practise a modified form of it.

First Steps in Training

Everyone knows that on taking a basket of young birds for their first toss and liberating them perhaps only a mile or so from the loft, they circle and turn first one way and then another and disappear in one direction, only to come back and go off in another direction. Much of this circling is the result of habitually circling their loft and the rest is an expression of the sheer joy of living. It is only when they have flown their fill that they make tracks for the loft. Many think that this tearing around the country is a good thing and teaches the young birds the countryside. However, it can lead to the formation of an unnecessary and bad habit. One method of counteracting this habit of circling around the countryside on liberation at the shorter distances is as follows. Let your youngsters out for their exercise as usual, and when they have flown their fill, get them into the loft without feeding them: a little seed thrown down will do the trick. The next step is to basket them and take them for a toss of a few miles. On their return to the loft, feed them in with their usual evening feed. Repeat this on successive evenings and soon you will find that they clear smartly from the liberation point to the loft and trap immediately. Youngsters treated in this way come to associate being liberated from the basket and flying back to the loft with being fed. It is an advantage during this training period to feed only once per day. Pigeons, in the opinion of animal psychologists, may not have a high IQ, but if they are repeatedly treated in the same manner they come quickly to associate one thing with another. For example, some of the young pigeons trained as previously related got to the stage of entering the basket of their own free will, evidently associating being in the basket with being liberated and fed.

Incentives to Return

In the early months of a young pigeon's life the incentives or the stimuli which induce them to return to the loft from being liberated are, basically, the desire for the familiar place—the loft, in which all their interests are centred—food, water and a perch. The first of these is the only one over which we, their owners, have any control and every use should be made of it. There is no need to starve youngsters. It is merely a matter of timing to ensure that they are hungry and due to be fed at the time they are basketed for a training toss. If this is repeated often enough they will quickly learn what awaits them on their return. This system of training must be continuous. Distance does not matter for the aim is to get the youngsters to race home to be fed. Once this lesson is absorbed we are on the way to producing from among them a good reliable racer or two. This early training, the object of which is to teach them to get off the mark from the basket quickly, is best kept to comparatively short distances so that no other problems are present in the form of changes in the weather or meeting other training batches. Once it is apparent that they have got the general idea and are coming more or less direct, which means between a minute and a half and two minutes per mile flown, you can increase the distance. How far you send them will depend on the topography of the immediate countryside and the facilities available for training. A good guide is that young birds cannot be sent often enough to the point where it is thought they should break from the main convoy in order to win. Familiarity with the direct line from this point is an essential to their training and will stand them in good stead at critical moments later in their life.

Revisiting Known Locations

It is a common experience that pigeons, when liberated, will repeatedly come back over the spot from which they have been liberated, and the opinion has been expressed that this is because they failed to get the proper line for home in the first instance and so back they come to try again. This may be so but to me there seem to be other reasons for this characteristic. I have noted over and over again how quickly pigeons can fix a location in their memory and be able to return to it, apparently at will. I have a considerable amount of evidence of how pigeons, liberated at one point, have, before returning back to their loft, actually visited another point at which some time previously they had been liberated. My attention was first drawn to this trait when visiting a friend. I took my young birds with me and liberated them at his house. Some days later he told me that my young bird team had come over his place and circled it several times before going off again. At the time I thought it was a coincidence and that he had seen them returning from a training toss, but when it happened again I was puzzled.

During one phase of my Army service I had several teams of young birds at various stages of training. Owing to the lack of transport their liberation points were fixed, with the result that their training route formed a letter Z from the loft to the furthest liberation point. To my surprise on the day that the most advanced team was tossed at the furthest liberation point for the first time, it was observed that they went over and circled their old liberation points. This meant that instead of taking the direct route these pigeons had followed the dog's leg route back over their old training points. When I heard this

I had men posted at all the old liberation points and instructed them to watch out for these pigeons. Their observation revealed that it took a week of daily flying from the furthest point before they more or less cut out flying the dog's leg route and started flying direct.

This tendency to revisit known locations must always be borne in mind when training youngsters. I have had several instances of it happening since my Army experience taught me to look out for it. Indeed, as recently as 1959, a team of young birds which received four tosses in a town approximately twelve miles north-east of their loft reminded me of this tendency. One evening these youngsters were taken twenty miles south-west and when an hour and a half had passed from the agreed time of liberation, fears arose as to what might have happened. After two hours things began to look serious and, moreover, it was getting late. Just then a batch of about fifty pigeons came over from the north-east racing hard and out of them dropped the eagerly awaited youngsters. As they settled on the loft it was seen that there were strangers among them, one of whom entered the loft. On examination, the stranger was found to bear the ring of the federation in which the fanciers of the town to the north-east competed. While this is not absolute proof that these youngsters actually visited this town to the north-east, the fact that they took over two hours for the journey and had swept up some other fanciers' young birds suggests that they actually did. The fact that pigeons can and do fix a location very quickly will be in the experience of most fanciers.

Strays

A stray enters your loft, you feed and water it, next day you take it away and liberate several miles from your loft and when you get back to the loft there it is waiting for you. You may do this several times, then one day it is gone. My explanation of this behaviour is as follows. As the average stray is confused, tired, hungry and thirsty, rest, food and water are its most pressing needs and desires. It finds them in your loft but the effects of fatigue, hunger and thirst are not thrown off in a single day, therefore, remembering a pigeon's inherent faculty of returning to the place in which its immediate interests are served, there is no need to be surprised that the stray returns to your loft. Finally, one day when it has recovered its confidence something jogs its memory and off it goes. Pigeons are equipped by nature to search for their food and having found it to be able to return, irrespective of the distance between their perch and the feeding place. Racing pigeons have a concentrated incentive to do so, since all their interests are centred on their loft.

Single Up

Once the youngsters have been taught to return to the loft in reasonable time from the training tosses the next step is to persuade them to fly alone or in the company of only one other pigeon. Liberating them in twos and threes pays dividends in this respect. Pigeons are by nature gregarious and will instinctively stay with other pigeons. However, no pigeon is likely to be an outstanding winner if it stays with the crowd; it must break somewhere. Pigeons like what they have been used to, and if your youngsters keep coming in a batch they are likely from sheer habit to stay with a batch instead of breaking early. If, however, they had some experience of flying alone or in the company of only one other pigeon, they will more readily be prepared to break away at a critical moment when racing. I think that this is one of the advantages fanciers with small teams have over fanciers with large teams, which are exercised and trained in considerable numbers. These birds are more likely to stay with the crowd and be influenced to a greater extent by the drag of numbers. One of the objections to liberating pigeons singly or in couples or trios is the time it takes, and the uncertainty that after all the trouble you have taken, the birds will not eventually all join up and home together. A little thought on the habits of liberated pigeons will meet this objection. The reason why they are liable to join up, even if liberated separately at intervals, is their habit of returning to the liberation point after we have been fairly certain they have cleared away. The solution is to liberate each pigeon or batch at different liberation points. If you have a car this is simple. Start liberating one or two birds and then go on for half a mile or so and liberate some more. Keep doing this until all the birds are liberated. The result is to give them all different liberation points and the time taken in moving from one liberation point to another separates them. You must use your own judgement as to the time and distance between each liberation.

Self Reliance

Some fanciers make a practice of sending up an old bird or two with their youngsters, with the idea that the experienced birds will guide the youngsters home or, as they say, give them a line. Like most other ideas, I have tried this one but am doubtful of its usefulness for two reasons. First, I want to train my young birds to rely on themselves. Second, I have always feared that the old birds may teach the young some bad habit, which in my experience is more likely than that they will instil in them a desirable one. I prefer to tackle the training of each year's batch of young birds with the avowed intention of making a better job of it than ever before. My method is to avoid previous errors of judgement, knowing that practically all the fault lies with me and my thinking and not with the pigeons. My youngsters are strictly segregated from the old birds, and even if I sent some old birds with the young ones I doubt whether they would home together. Moreover, if they did it would be the old birds which would come back with the young, from sheer force of numbers. On race days, when perhaps every youngster in the loft would be away racing, I have tried putting a few old birds in the trapping compartment to encourage the returning race birds to trap. However, it fails to work with the young racer immediately it recognizes that there is something unusual happening. They trap better without the old birds, even though it sometimes takes quite a time. It should also be remembered, with regard to young birds trapping from a race, that apart from their understandably excited condition, it is perhaps the first time they have found themselves out on the trap alone; therefore in their experience it is something unusual. This is where the ingrained environmental habit of trapping immediately day after day pays dividends.

The Doubtful

Another reason I have for training my youngsters separately is that I want to find out the workers among them. At one time I would deliberately toss two good reliable pigeons with a doubtful one in the hope that the two would improve the performance of the other by example. Now, as far as possible I toss two reliable pigeons together and the doubtful one on its own. One thing that is certain is that pigeons pick up bad habits much quicker than you can instil desirable habits into them. Do not blame the pigeons for this state of affairs. In the majority of cases it is either the environment or the line of thought and management that is unsuitable. I recall that while training a batch of youngsters using the method outlined, we found that invariably two or three pigeons were always among the first to arrive back to the loft. Armed with this information we then allowed these few selected birds to be liberated on their own, free of the others. We were delighted to find that this improved their times considerably. Eventually, we had three teams: the very good, the good and the not so good. In fact, we worked a relegation and promotion system. This and other observation convinces me that the indifferent performer is a real handicap to the others in a team of young birds under training. I am convinced that young birds should be trained in groups. Divide them how you like: by age, colour, or sex so long as each group does not exceed a dozen. By this means your whole team is unlikely to be victim to the same unfortunate experience and, as already discussed, there are advantages in their learning to fly in small groups. When flying, pigeons are certainly democratic. There appears to be no leader, they turn and twist, rise and swoop at the will of the majority. It is a well-known fact that one or two old pigeons in a team at exercise slow down the pace of the others.

Owners' Behaviour

Readers might have noticed how differently fanciers behave on the arrival of a bird from a race from the way they behave when calling their birds in from exercise daily. Not only does the owner behave differently, but often the whole set-up is abnormal. The children are kept indoors, the dog is locked up in case he barks, and there is no washing on the clothes line. Yet these things, the playing children, the barking dog and the flapping clothes, all form part of the everyday environment of the pigeons. No wonder the returning racer is nervous and hesitates to trap, disliking, as all pigeons do, anything unfamiliar. To make matters worse, if their owner normally whistles to call them in, he now finds his lips dry and his whistle rather reedy by comparison. Instead of moving about normally, he slinks around like a thief, hiding in some corner ready to pounce on the bird when it comes in. Can we be surprised if a pigeon after one experience of this sort of treatment shies at entering the loft? The inside of the loft is often different from normal. The water fountain has been moved out of the way or the timing clock is in a prominent position. A high-strung pigeon notices all the things that are different. Moreover, it has been proved by experiment that pigeons are more sensitive to the position of an object than to its size, shape or colour. For instance, if the nest bowl is normally in the right-hand corner of the nest box and it is moved to the left-hand side, the pigeons will notice it immediately and may refuse to enter the nest box, at least at first. However, substitute a larger or smaller nest bowl for the original, and this will not be noticed so quickly. It may produce a little hesitancy, as to a lesser degree will a change of shape or colour. Just alter the position of the entrance to a nest box and watch how long it takes the inmates to become accustomed to it, to the point when they will fly up to it without a lot of neck stretching, head bobbing and peering as a preliminary to the act. Alter the colour and they will scarcely notice it. Remember also that they are equally as quick at noticing that something is missing and so do not alter your loft around, particularly on race days.

PART TWO: THE OWNER AND HIS PIGEONS

Team-work

Successful pigeon racing consists of team-work between the owner and his pigeons and good team-work is the result of constant practice between all the members of the team. In this case the members of the team are the owner and his pigeons, each doing their part over and over again in order to reach perfection. The natural instincts and habits of pigeons make them suitable subjects to perform in a given manner in certain circumstances, provided they get constant practice. It behoves the owner to do the thinking and planning and then to train his birds to carry out his plans. The plans must of course be logical and practical, otherwise they are useless. They must also be carried out with persistence and patience, if they are to succeed. The importance of trapping cannot be over-emphasized However, judging from the performances I have witnessed, on many occasions little or no thought can have been given to the subject and much less to producing a carefully prepared plan which has been assiduously practised. One of the most amusing attempts at trapping I ever witnessed occurred many years ago. I happened to be at another fancier's loft when his first bird arrived and kept circling overhead. The owner stood in front of the loft rattling some peas in a can and whistling, or at least trying to do so. Eventually, the bird landed at the back of the loft and round the fancier went to chase it to the front where the trap was located. The bird came round to the front but shied off again when the owner came round rattling the can of peas. The fancier then decided to go inside the loft, from which came the noise of much scrambling and fluttering. When this had died down somewhat, the returned racer, who had got over some of its fear, came forward hesitantly and peered into the trap. At that very moment the owner called to me to bring the clock, which he had left under the garden seat where we had been sitting. I picked up the clock, but waited until the bird had dropped through the trap before opening the door to hand it to the owner. All the time the fancier was in the loft he kept shouting, asking where the bird was and what it was doing. I opened the door of the loft just as the owner made a grab at the pigeon, which he missed. As all the other birds were scrambling around he had to wait until they were settled before he could locate the bird he wanted. When he spotted it, he made another lunge to catch it and in doing so his gold watch chain became entangled in a wire contraption on the top of the water fountain. When he straightened up, the fountain was dangling between his legs and water was being splashed all over the place. The loft was in pandemonium as the owner slipped and slid on the wet floor. When the bird was finally caught, I discovered why the fancier could not whistle: he had the timing thimble in his mouth all this time.

Practice

Trapping and the catching of pigeons on their return from a race should be a planned affair, in which each member of the team knows what to do and can depend on the other member doing his part. Such a standard can only be obtained by daily practice throughout the year. The simplest plans are always the best. I have seen all sorts of trapping arrangements and gadgets, many of them ingenious, and it was a friend of mine who produced the best by always feeding his birds in the one corner of the loft which was in full view of any bird on the trap. Every day at the end of his birds' exercise he stood in the middle of the loft in full view of any bird in the trap and threw down food into the corner. On race days he behaved in exactly the same way and his return racers, without any hesitation, dropped in and joined the other birds feeding in the corner and were easily picked up from among the crowd. His clock stood on a little shelf out of sight and the timing thimbles were in a slot in the shelf. Slipping the ring off the bird's leg, he lifted a thimble and in a matter of seconds the bird was timed. The whole affair was just a matter of practice and usage: the pigeons expected the owner to be there and he knew where they would go on entering the loft. I always admired the skill with which he threw down the last handful of seed in such a way that the new arrival was instantly surrounded by the other birds, easily picked up to remove the ring and put down all in one quick movement.

Careless Thought

The design of your loft has much to do with the methods you adopt, and the badly-designed trap in relationship to the rest of the loft can be a great handicap. A fancier once complained to me about the trapping arrangement of one of the ex-Army lofts he had bought, declaring his birds refused to trap on race days. The difficulty was that he could not reach the far corners of the trap, so the bird which trapped easily eluded his efforts to catch it. But much worse was his practice during the week of opening the windows and allowing his birds to go out and in through them. In contrast, he was expecting his birds to come in on race days through the trap.

Get Them In

Pigeons should never be caught anywhere near their point of entry into the loft or they will hesitate at entering, since, if it is repeated, they come to associate one thing with another. The first aim in trapping is to get them to enter without hesitation which is only achieved, dare I repeat, by constant practice, every day of the year. Once inside the rest is up to you; the important thing is to have your plan carefully worked out, trying it out beforehand. The next thing to remember is the timing clock with the thimbles near at hand. Finally, there is always the unexpected. A friend who is of the good and simple trapping school, laughingly recounted how he was hoodwinked by a returned racer. He was sitting in the loft just looking at his birds when he became aware of a pigeon alighting on the trap. Without rising he reached for the seed he used for trapping. As he threw it down, the bird on the trap dropped in and at the same time the other inmates flew down from the perches into the corner. My friend then rose to get hold of the racer from among the other birds and to his surprise it could not be seen. Taken aback he had another look, but it was not there. His first thought was that it had not come in, and he was on the point of going outside to look, when he spotted it on the floor behind him. What had happened was that his bird, instead of going to the usual corner for food had, in the confusion of the others flying down from their perches, slipped behind my friend to the water-fountain for a drink.

Use Their Legs

Encourage your birds to use their legs instead of their wings in the loft. Divide the loft into small compartments with small apertures at the bottom through which the pigeons can walk from one compartment to the other. The compartment doors should be kept closed, thus preventing the birds from attempting to fly about the loft. One trap for the whole loft is much better than, for example, one for each compartment.

Traps and Trapping

It is quite easy to arrange that the birds trap into a central compartment which gives access to all the others. Always feed your birds in the trapping compartment after exercise so that they will behave normally on race days. In my loft at Old Neilsland the trapping compartment floor was about nine inches higher than the floor of the rest of the loft. Normally, the birds dropped through bob-wires into their respective compartments. On race days these bob-wires were fixed and I just picked up the race birds as they tried to go through them. With the clock and thimbles handy, I have timed three birds single-handed within sixty seconds. The advantage of this type of trapping is that you are in the compartment when the birds come in from exercise, and when they have finished eating they slip through the free-swinging bob-wires into their compartment. On race days you are, as usual, in the compartment with a few hungry birds at your feet and, of course, no perches. The racer drops inside on seeing the others eating and it is a simple matter to get behind the bird you want to catch. Naturally, it attempts to elude you by going through the bob-wires but with its head through the wires it cannot see you and is easily picked up. As has already been said, it is all a matter of what is best in the circumstances in which you find yourself. I have had many different kinds of traps: round holes in the roof, horse-stall partitions, the ordinary let-board, and ordinary bob-wires. The latter, properly hung, are the safest and best and have the advantage of being easy to fit with a warning bell for us in the longer races. The trap at Old Neilsland could be used as the open door trap and was so used in the shorter races. In the longer races the birds came in through bob-wires to which a warning bell was fitted. In order that the birds should become used to both methods of trapping it was my practice to use the open doors and the closed doors with the bob-wires on alternative days. One important thing in deciding what kind of trapping system to employ is the question of the drop. If the drop is of necessity a short one it should not exceed nine inches, as this is about the maximum distance a pigeon will drop unhesitatingly without using its wings. If the drop requires a pigeon to use its wings, then it should be at least four feet; anything under this makes a bird hesitate. These are just a few general observations which may prove useful; the main thing is that your birds should be made to trap quickly every day in the way you would like them to do on race days.

Train Yourself

Pigeons should be either in the air or in the loft. I know how pleasant it is on a nice day or in the evening to sit outside the loft with your pigeons picking around and evidently enjoying themselves. On the odd occasion, this may be all right with experienced old birds in which habits have become fixed. However, since young birds are too easily spoiled and because it is very difficult to get rid of a habit once it is formed, young birds should never be treated in this way. Never be discouraged if by reason of circumstances your trapping arrangements are not to your liking and you cannot alter them. Get busy training your young birds to make the best of the circumstances and remember it is up to you, their owner, to do the scheming. You must train yourself to behave regularly over a long period of time with the result that you can become more certain of the behaviour of your pigeons. Move about the loft confidently and with purpose and your birds will soon get used to your movements. There is no need to be on tiptoe, like a cat who's stalking a bird. I am reminded of a visit I paid when young to one of the pioneers of the sport in Scotland, both as an active racer and official. When I arrived he was at the foot of his garden and his loft was between him and me. When he saw me he signalled me to stay where I was, and pointed to his youngsters on the top of his loft between him and me. He then started coming very slowly towards me and when near to the loft he got down on all fours and started crawling past the loft. This startled his youngsters, and the dog next door, seeing him through the hedge, started barking, no doubt at his strange behaviour. Up went a few of the youngsters, fluttering around as my young pigeons do when first on the wing. I am afraid my visit was not welcomed. I did not say anything: was he not a great fancier and I only a beginner? I have never forgotten the incident and its obvious lesson. Behave naturally at all times for your pattern of behaviour will be reflected in the habitual behaviour of your pigeons. Most things which go wrong are our own fault; the careless action, the faulty reasoning and the lack of understanding of what our pigeons may be feeling.

Starting Something New

Pigeons are very much creatures of habits, good or bad, which once formed are exceedingly difficult to alter. Even when you think you have broken a pigeon of some undesirable habit, it will recur at the crucial moment, to our discomfort. Always start anything new with young birds; they know no different and are very adaptable. Older birds like what they have been used to and resent change, especially sudden changes. They take longer to learn something new and are rarely 100 per cent reliable in it. Work out your plan for trapping, practise it constantly and it will reward you well.

Foolish Things

I do not think there is a foolish thing a pigeon fancier can do that I have not done, or that there is a fad or fancy that I have not at one time or another believed in and practised to my cost. Many of these fads and fancies, perhaps cults is the word are hard to kill, and the fact that successful fanciers have believed in and practised them give them an authority they do not deserve. Strangely enough, many of us are doing the right thing for the wrong reason, and attribute virtues to practices which have no effect whatsoever on the ultimate result. In retrospect I marvel how my own pigeons have won at times, despite my mismanagement of them. So many of our wins, of which we are so proud and for which we accept the credit, are fortuitous affairs brought about by circumstances in which we, the owners, took no part in so far as planning and preparation are concerned. As a club secretary I notice many things. For example, the fancier who regularly picks and pools the first bird in his loft: unless it is actually a winner it passes unnoticed by the other members. I also come to know each member's good pigeons from their ring numbers published in the club result. I also see the men who repeatedly fail to pick the right one; the reason why is their problem but too many pass these incidents off as just bad luck. Similarly, when another pulls off a winner in all pools it is just that he has been lucky. It is to the reasons for our failures we should give all our attention, remembering that the easy answer is not always the correct one.

Training Ourselves

We must train ourselves to observe and, equally important, to put the right interpretation on what we observe: to differentiate between the likenesses and differences which exist to trap the unwary. Observe and plan, taking care your plans are practicable, and that you can carry them out, otherwise you are simply indulging in wishful thinking. There may be those who think I have overemphasized the value of planning and practising. The following story may dispel any doubts.

A few years ago I met an old acquaintance who, after many years in the sport, had just experienced two years of very successful racing. Laughingly he told me that he could not understand his success. He was away from home from Monday morning to Friday night every week and so he only saw his pigeons on Saturday and Sunday. However, he was racing very well, in fact, better than he had ever done before. I questioned him very closely and found that in the circumstances he had been forced to do a great deal of detailed planning, in order to leave written instructions for his wife to carry out in his absence. He had been really worried at first and even contemplated changing his employment. His success had changed his ideas both of pigeon racing and his wife. He has found in her an able and willing partner, doing the pigeon chores with enthusiasm and interest. Under different circumstances he would have never allowed her near the loft, fearing she would make a mess of things. She, in her turn, would have grumbled, complaining the pigeons were not in her sphere. Now they are a happy and successful partnership and are agreed that two heads are better than one. He confessed to me that he had never before given such deep and detailed thought to the subject and that what he had first thought was an insurmountable difficulty had turned out to be a blessing in disguise. With great glee he told me his birds were fed and exercised to the exact minute and pea, and any which stayed out were neither coaxed in nor fed when they did enter the loft. His parting shot was that we are all too soft-hearted. I repeat we all need to train ourselves to observe and, equally as important, to put the correct interpretation on what we observe, to differentiate between likenesses and differences which exist to trap the unwary. Observe and plan, taking care that your plans are practicable and that you can carry them out. Remember, with Burns:

The best laid scheme o' mice an' men
Gang aft a-gley
An' lea'e us nought but grief and pain
for promised joy.

We have discussed the training of young pigeons and the training of the owner and his pigeons into a well-drilled team. We will discuss the training of old birds in its fullest sense, with the emphasis on yearlings.

PART THREE: MATURE BIRDS

Stimulus of the Mating Cycle

You will recall that the incentives which influenced young birds to the loft were the security of the familiar place and perch, the company of other pigeons, and food and water. With the coming of the breeding season other feelings and emotions are brought into play. The early signs of the desire to mate are the tendency of hens to mate together, the odd pigeon lost the previous year suddenly turning up, and, of course, the general restlessness of all the inmates of the loft. Hens, in particular, are inclined to be unsettled at this time. Each phase of the mating cycle as it occurs in its regular sequence is brought about by changes controlled by the endocrine glands. This glandular activity, in addition to bringing about the behaviour of pigeons during the changing phases of the mating cycle, stimulates and quickens at this time all the body processes. The result is that as a rule pigeons are found to perform better when racing during this period. Our scientific friends may ask what proof we have that it is the driving, sitting or feeding conditions which urge pigeons to race home during these phases of the mating cycle. May it not simply be the super physical fitness and form brought about by the more than usual active functioning of their glands? Personally I am inclined to think it is a combination of both, plus the fact that the distance, pace of the race and other conditions are reasonably favourable to them on that day. Need I fear any contradiction when I say that pigeons do not always respond to a nesting condition in which they have on a previous occasion scored, nor for that matter do pigeons respond which look and give every outward signs of form? On the other hand, when they do score they are usually in one of the favoured nesting conditions, and the fact that they scored is proof of their good form.

Winning Conditions

In my book *Pigeon Racing—Win with Olympic*, Chapter 10 gives the result of an analysis of the performances of between

600 and 1,000 pigeons flown on the natural system of which detailed records were kept. The object of the survey was to find what phase of the mating cycle gave the best average results. Hatching topped the list, producing the first bird to the loft out of every seven sent in this condition. This category included pigeons sitting over sixteen days under which a small youngster was slipped for an hour or more prior to basketing. Feeding a small youngster (young up to nine days old) came next, producing the first bird to the loft out of every nine sent in this condition. There is a decided tapering off with youngsters ten days old or more. Driving produced the first bird to the loft out of every nine sent driving. The next condition, sitting on eggs, i.e., eighteen days of incubation, was divided into three periods of six days each. The periods one to six and seven to twelve were about equal, producing the first bird to the loft out of every eighteen sent. Cocks showed a preference for the first six or seven days and definitely tapered off after ten days though the hens showed a tendency to keep their form for a few days longer. Birds sitting from the thirteenth day to the time of hatching only produced the first bird to the loft out of every seventy-two sent. Unmated pigeons in the loft and sent racing produced the first birds to the loft our of every twenty-five sent. Birds feeding young over ten days old were the least successful. In connection with this point, quite a number of cocks are said to score when sent feeding a big youngster; in fact, they are actually driving their hen to nest or sitting on the next round of eggs for a day or two.

Facts Versus Opinion

The above results are factual as opposed to opinion and I readily admit that my opinion, based on what I had read and been told, differed from the above facts. Every fancier has his favourite condition in which to send out his pigeons and by which he swears. It was this clash of opinions which prompted me to try to find the actual facts. In the years which have passed since this analysis, I have had ample evidence of the accuracy of the above placings. In every case the days count from the day the first egg is laid to the day of basketing.

Accurate Records and Notes

Fanciers are on the whole very careless in keeping records, and many sales are spoiled by the lack of essential detail. For several years I have made it a practice to enquire in what nesting condition pigeons are being sent racing to open events and the Nationals. The favourite one seems to be sitting on eggs for ten days, yet when pressed for the exact date of laying I have found all too often that the actual sitting period varied from five or six to as many as fifteen days, and not ten. Similarly, when visiting a loft to see a winner I meet with the same inaccuracies. Pigeons which, you are told, were sent feeding a big youngster, were obviously sent—from the size of the youngster still in the nest—when the youngster must have been about a week old. These experiences make one very sceptical when seeking facts. In preparing a pigeon to repeat a previous performance, surely every effort should be made to send it in exactly the same condition as before, and I am of the opinion that a day or two either way may make all the difference. It is true that a really good pigeon will usually be among the leaders, but in present-day competition, the very best is required to win. The answer to all this is to keep careful notes.

Soft Food

From my experience and the analysis of the data which formed the basis of the factual investigation described above, it appears to me that the presence of the soft food and the consequent desire to feed is the most lasting and potent of all the phases of the mating cycle. There are those who think that the presence of soft food in the crop of a pigeon sent racing will turn sour in the crop, to the detriment of the bird concerned. This line of thought is purely hypothetical, based on the fact that mammals from which their young have been taken away suffer a great deal of discomfort for a period. Close observation of pigeons from whom the young had been taken away in or about the first day of hatching disclosed no ill effects. Apparently the soft food is digested and absorbed by the parents themselves. In that case, the racer who has to spend a night out and has soft food in its crop which it digests and from which it derives nutriment is much better sustained to continue the next day than the bird with the empty crop.

Driving and Sitting

Driving produces good results, but much depends on the management of driving cocks if the best results are to be obtained. The stimulus here is similar to that which produces racing urge and form in widowhood cocks: the stimulus must be properly directed. The most popular condition, based on opinion, is sitting on eggs. This condition lasts the longest, roughly seventeen days. The majority of birds are sent in this condition and from sheer force of numbers and by the law of averages should provide most winners.

Other Conditions

The other conditions are best avoided; I would rather wait and jump a pigeon a considerable distance than send it to a race without some special incentive. Broodiness is more liable to wear off than the desire to feed, especially while the soft food is still being produced. The length of time broodiness or the desire to feed will last while the birds are in the basket will depend on how long they have been active before basketing. It is a fact that the best results have been obtained, on the average, by birds in the early stages of feeding or sitting, i.e., the first seven days. It has been noted that many birds sent sitting refuse to take to their eggs on their return, while it is rare to find a bird refuse to take to its young. This is significant.

Irrespective of what condition, in your opinion, is best, I do not think that anyone will be found who does not agree that during the changing phases of the mating cycle racing pigeons are found to make that extra effort so necessary to win and to react more strongly to one phase than the others. A natural reaction is to ask why. No doubt the activity of their glands will have something to do with it. Moreover, we can never discount the possession of a physique capable of sustaining the purely physical effort required. My experience in studying this problem leads me to think that the good pigeons are those which have come to learn from previous experience that to satisfy their pressing physical desires to sit, feed or eat and drink they must return to their loft. In orthodox training the fact that a few pigeons come to learn to race home appears to me to be purely accidental. If this is so, there must be many other mediocre performers who, with thoughtful management, could be taught to race back to their loft.

Training Yearlings

Our next topic is the training of yearlings who have come under the influence of the mating cycle for the first time. Since they have had no previous experience, it gives us, their owners, the opportunity to inculcate in them the habits we desire. The tendency is to start racing too early, with the result that some basket training on the road is indulged in while the birds are on the first nest of the season. I have noted what fanciers do during these early tosses. The man who gets up early in the morning will tell you he has had his cocks out to twenty, thirty or forty miles. When you ask him about the hens he says, 'Oh, the hens are sitting when I send the birds away in the morning, but I will get the hens tossed on Saturday and Sunday afternoons when the cocks will be sitting.' Now this seems quite logical, but in teaching pigeons to race home it is far from logical. We send our pigeons on these training tosses in order that they may practise the use of their homing faculty, for physical training and to become acquainted with the immediate countryside between the liberation point and the loft. Well aware that pigeons race better when sitting or feeding, how many pay the slightest attention to this aspect of their training? We know that cocks sit roughly between the hours of 9 a.m. and 4 or 5 p.m., and the hens for the balance of the twenty-four hours. I think we are justified in thinking that the sexes must experience the emotion, feeling or whatever it is which makes them sit, more strongly between these hours. It seems logical to me that if we wish to teach these pigeons that they must race home in order to satisfy their physical desires, then the time to send them is when we think their emotions, feelings and desires are being felt the strongest. The logical time to send cocks is when they are due to take their turn on the nest, i.e. late morning or early afternoon, and the hens in the early morning or early evening.

They Have to be Taught

I am convinced that pigeons have to be taught these things, just as they have to be taught to trap, and eventually do so from habit. They can also be taught to race home if handled methodically on a well-thought-out plan which is continually practised day by day. Then they will come to do it from habit. If it were possible to toss our pigeons every day, there is no doubt they would soon respond, but there are very few of us who have the time or facilities. We got splendid results during the last war when it was possible to toss pigeons two or three times each day. One thing that has to be remembered is that you must persevere. As Captain Roland Smith used to say 'You cannot expect to learn to play the piano in three lessons.'

Pigeons can be taught to appreciate their nest, eggs, young and mate by the partial separation of the sexes during the racing season. All that is required is two separate compartments for the breeding pairs, one containing nest boxes only and the other with only perches in it. During the period when the hens are normally sitting the cocks are kept in the compartment with the perches. In the morning the cocks are exercised, and when fed in they are let into the nest box compartment and the hens removed to the perch compartment. In the evening the hens are put out for exercise and fed in, after which they are allowed through to the nest box compartment. The cocks are then put out for exercise, fed in and allowed a little time in the nest box compartment beside their mates and later put through to the perch compartment for the night. This system teaches the pigeons that the only place they will find their mate is in the nest, and has the effect of making them keener to sit and to feed and improves trapping. Above all, it gives them a keener appreciation of all the things they desire and increases the chances of a greater number of them learning to race home in order to satisfy these desires. This is not a new idea by any means and is being practised with success. In fact, it was used by a very successful fancier and National winner of very many years ago. His idea, however, was that it delayed hens laying too soon and prevented driving cocks from running themselves and their hens to a shadow. To me the logic of it is that there is not much use in a pigeon desiring its mate or to sit or feed if it has no clear perception of what it must do to satisfy its needs. Pigeons sit on their eggs and feed their young because it satisfies an emotion or physical urge to do so. They are not motivated by an anxiety nor have they any notion of the ultimate result of what they are doing. It is, therefore, a fallacy to think that they race home from anxiety over whether their eggs or young are being looked after during their absence in the race basket. In fact, they are possessed by an intense physical desire to sit or feed, and they race home to satisfy that desire. In just the same way if they are hungry they seek food to satisfy their hunger or when thirsty they seek water to drink. Similarly, the bird that has produced soft food wants to regurgitate it, an action which not only satisfies it but also its young. As far as the parent bird is concerned this action is not prompted by anxiety for the present and future welfare of its young, but, for all we know at present, simply to get relief from what is perhaps an uncomfortable physical feeling. Do not think that I am ignoring the parental instinct which is complex and would appear to form a combination of all the others; however, a discussion of that is beyond the scope of this book.

Driving Cocks

During my early days in the sport I was taught, and believed, that it was dangerous to send driving cocks racing, especially yearlings. Moreover, when you think about it, yearlings are naturally, from their inexperience, more liable to be lost than older birds, no matter in what nesting condition they are sent racing. However, experience teaches us otherwise. When, in a very hard race from almost 400 miles, with only a handful of pigeons home on the day, the first two belonged to the same fancier, I decided to visit him on the Sunday after the race. I was astonished to find the successful birds were two yearling cocks sent driving, which made me think again. Driving cocks are no different from others; they have to be taught and the actual time they are driving is short. The golden rule with driving cocks is never to let them out of the loft with their hens. Use every artifice you can think of to instil into them that the only place they will find their hen is in the loft. The old adage, 'You never miss the water until the well runs dry' applies very much to pigeons. They take so much for granted that it is only when they are deprived of something they show any real appreciation of it. This is the crux of the matter and something over which we, their owners, have a measure of control and can, by intelligent use, turn many things to our advantage. Time your pigeons to be hungry when you let them out for exercise and you develop the habit of good quick trapping. Time them to be hungry and due to be sitting or feeding when you send them on a training toss and the chances are that the majority will learn to race home. Take a driving cock away from his hen, put him in a basket away from the loft, then later either give him a short toss or just liberate him outside the loft. Alternatively, or in addition, just throw him outside the loft by himself. At first, he will fly around before coming in, but if repeated time after time he will soon come to the stage of just wheeling round and entering the loft. This is evidence that he has learned where he will find his hen and indicates that more than likely he will react similarly when sent racing. This is no mere hypothesis, but the result of practical experience and it works. I followed the practice of taking all driving cocks to the office (a distance of ten miles) every morning and liberating them singly. I once told a fancier that he should never let a driving cock out with his hen and that they should be taught that the only place they would find their hen was in the nest box. Later he startled me somewhat by telling me that he now locked hens that were being driven in their nest boxes so that the cocks would always find them there. What he had overlooked was that a driving cock is perfectly happy when the hen is in the nest box. On the same principle as the driving cock and going a stage further, it is a good idea to exercise the sexes separately so that they are only together inside the loft. This produces a complete concentration of all a pigeon's interests in the loft, and nowhere else.

On the question of the number and distance of tosses. With yearlings they cannot have too many but the distance need not exceed twenty or thirty miles. The object is to get them to race back and once you get them doing that it is time enough to think about increasing the distance. The older birds need not be harassed with these early tosses; they are likely to be set in their habits. Furthermore, they are the survivors of many reared and for that reason presumably above the average. The yearlings have yet to learn and should be given every opportunity to do so. Give them something easy to do and repeat it often, preferably at the breaking point. This may not be where you think it is, but enquiries from others may give you a clue. Getting that extra distance is a great temptation during training; just let them get a bad day and all your carefully-thought-out plans and preparations are upset—birds out for a night or two, eggs forsaken, matings broken up and all to no purpose. It is team-work which counts; we do the thinking and planning and then we train ourselves and our pigeons to carry them out. Keep thinking again and remember that practice makes perfect. What I have written about training in this chapter is only a very sketchy outline of a vast unexplored field which in my view holds great possibilities to those who care to explore it further.

5 The Signs

Physical Features

No book on racing pigeons would be complete without some reference to the visible physical features popularly supposed to be possessed by really good racing pigeons. The consensus of opinion favours a pigeon symmetical of form which, when in the hand, confirms this impression of balance. Each of us has our own views on minor details and since there is no standard, as is usual with the fancy varieties, it is surprisingly how well-defined the ideal racing pigeon is in the minds of the majority of owners. For exhibition purposes bloom and condition are primary essentials and the individual bird must be free from structural faults or the exaggeration of any physical feature or abnormality of feather. The success in the show pen of even the very best of pigeons is dependent on the views of the judge on the little details. When it comes to racing, however, it is often said that they win in all shapes and sizes and I admit at times that this appears to be so. There are exceptions to most rules, and the unlikely pigeons which win the odd race are the exceptions to the general rule, and should be treated as such. All too often we are prone to be guided by the exception rather than the general rule or average.

Features in Common

The really good pigeons I have seen and handled have all had many things in common; perhaps not some of the things we have come to look for and expect. Of the things they had, first and foremost, was their evident good health, vigour and vitality. Moreover, in the majority of cases they came pleasantly to the hand, giving the impression of buoyancy and life. I have no intention of attempting to describe in detail the physical features I would look for and expect in the ideal racing pigeon since, as a beginner, I read many descriptions by others and they only added to my confusion. The recognition of reasonably good pigeons is something which comes from experience; it is something you have to learn for yourself. We know, of course, that the visible physical features are not the whole story. Recently, while in the company of a number of fanciers who were discussing the merits of a pigeon, I turned to an old and consistently successful fancier and asked him what he thought. He replied, 'It is the kind of heart and what is inside its head that matters.' Thus, we are thrown back on the unseen and indefinable, around which so much mystery has grown with its accompanying theories amounting at times to black magic. In our enthusiasm for our beloved pigeons we attribute to them virtues, physical, mental and even moral, many of which we would be hard put to substantiate in fact.

Facts and Fallacies

The separation of the facts from the fallacies is no easy task, especially as I have most probably believed in and practised many of them and the thoughts and beliefs of a lifetime are not so easy to discard. I know from experience how they can keep creeping back into our reasoning, in spite of our efforts to shut them out. When we come to try to reconcile practice with theory, on occasions they would appear to contradict each other, and we may be inclined to take one side or the other. On such occasions we should carefully think again, because very often it is our interpretation of one or the other that is at fault. The little I know about pigeons was learned the hard way, and in retrospect I have every cause to squirm at my own foolishness in the past. I, therefore, dislike to see others make and pay for the same blunders as I have committed. The thing which shocks me most is the apparent blindness to realities, to things going on under some fanciers' very noses which they appear not to notice, or if they do notice pay no attention to. They go on living in hope, indulging in wishful thinking; their odd success seemingly makes them forget their failures. I am reminded of overhearing a little group of Army Pigeon Service men talking. One, well known for boasting of his successes to anyone prepared to listen, was airing his views. Another, apparently becoming exasperated, interrupted him by saying, 'Your wins seem to worry you as you are always talking about them.' The boaster replied, 'What about your wins?' The answer was, 'My wins never worry me, it is the times I have lost and why I lost I keep thinking about.' It is not so easy to retain a realistic attitude to success and failure. Kipling knew this when he said:

If you can meet with Triumph and Disaster
and treat those two imposters just the same.

Homing Faculty

To return to the unseen and undefinable, let us see if we can sort them out. First must be this mysterious thing we call the homing faculty. Exactly how pigeons find their way home remains a mystery. There are many theories and we all have our own ideas about it. This we do know: whatever it is: racing pigeons possess it in a marked degree. We also know that racing pigeons do not all possess it to the same degree. Many otherwise well-bred and reared pigeons fail at comparatively short distances. It is reasonable to assume that they all do not fail from weakness in their homing faculty; physical inability and the other hazards to which pigeons are exposed take their toll. I have no doubt that it is of a complex character, subject to many internal and external influences at which at the moment we can only guess. The important thing is that it appears to be inheritable and develops with usage and training. In previous chapters the importance of early training, in order to develop to the full inherited characteristics, was emphasized and in this connection the homing faculty is no exception.

Stamina

Numerous superlatives are freely used to describe the merits of pigeons, but it is a different matter when we try to pin them down to realities. However, there is one outstanding characteristic possessed by all good pigeons and that is endurance or stamina. Again it is inherited and then developed by training. Some may doubt that inherited characteristics require to be developed, but there is nothing surer than that the individual pigeon is as much the result of its environment as its heredity. We, their owners, create the environment and if this thing we create does not encourage and develop the inherent characteristics then we have no one to blame but ourselves. When considering this matter it is as well to remember that only a proportion of racing pigeons possess homing ability and stamina to the degree of racing successfully from 500 miles or more. It is also our experience that a pigeon, by its performance on one occasion, may have given us the impression that it possessed these two characteristics yet on a subsequent occasion it has failed. In saying this, I am disregarding all the hazards to which pigeons are exposed when sent racing, such as wires, hawks, shooting and the many other fates that overtake the very best of pigeons at times.

Form

On the day a pigeon wins there is undoubtedly something more than homing ability and stamina required. Is it this unseen indefinable something we call form? Whatever it is, it varies from time to time and is present one week and gone the next. Physical ability and stamina are fairly constant, and except for the effects of old age, injury or disease, can be expected to be enjoyed for life. The changing phases of the mating cycle vary in their intensity in the individual and between one pigeon and another and, all other things being equal, may well be the factor which makes the difference between winning and losing between pigeons of equal physique, stamina and ability to home. The balance is evidently a delicate one, influenced by a number of factors which, when in harmony one with the other, produces this thing called form. We know that certain pigeons give a good account of themselves when in a particular nesting condition, others may respond to two or more nesting conditions. Our problem is that they do not always respond and the question is, what makes the difference? Bearing on this problem, but not altogether solving it, is the fact that while some pigeons will score at any distance, given the favoured nesting conditions, others will only score at the longer distances or when the race conditions demand long hours on the wing. When studying the data referred to in the latter part of Chapter 4, it revealed that top form came in cycles at roughly six- or seven-week intervals. It might last for two or even three weeks, fade away and then come again at the peak period, provided there had been no break in the normal sequence of the mating cycles. There is some evidence to suggest that pigeons rarely reach peak form until they have cast their first primary flight, which normally, in the majority of pigeons, takes place when they are sitting on their second round of eggs. In this connection I do not wish to be misunderstood or to mislead the reader. I do not imply that a pigeon will not win until after it has cast its first flight. Racing starts so early now that pigeons mated in the last week in February or the first week in March will still be holding their first flight when sent to their first one or two races. At a rough estimate I should think that only 5 or 6 per cent of the pigeons competing in these early races will have cast their first flight. With regard to pigeons which seem only to score at the longer distances, I have found that the majority of these are late moulters, i.e. pigeons which do not moult their first flight until sitting on the third round of eggs. It would appear, therefore, that the moulting of the flights has an influence on form and, for my part, any pigeon which holds a flight when there is no apparent reason for it doing so is suspect and I would rather wait and see what happens than send it racing.

Casting a Flight

The normal natural periods when pigeons are not likely to cast a flight are mating, cock driving, hen about to lay, and when both are feeding young. This leaves a period from about the sixth to the twelfth day of sitting when flights are likely to be dropped. Pigeons which are flown down, ailing or injured are unlikely to cast a flight until well on the way to recovery. The regular casting of the primary flights at the normal times is an indication that all is well, and if you care to keep a record it is surprising how regular the casting of each flight is year after year with birds mated early in March. I have tried mating birds later with the idea of getting them to the longer races with their second primary out and growing instead of their third. The result has been that some of them cast two flights (second and third). I am of the opinion that a pigeon in form with nine primary flights is a much better proposition than one which has been treated unnaturally and sent with a full wing (ten flights) and, moreover, I do not mind if it is the fourth primary that is missing.

The Finer Indication of Form

We have now discussed two known factors which together give us an indication of the likely presence of form, i.e. one of the favoured nesting conditions and the regular casting of primary flights. However, these are not the whole answer. A pigeon can have moulted and be in a favoured nesting condition and still lack the delicate balanced peak form which we seek. What then can we look for? There are one or two other visible signs which, when present, are up-to-the-minute indications of form: a keen healthy appetite, the tightening up of the feathers and the bird's general demeanour in the loft, which very often is merely a slight change of behaviour. Some birds become quiet at this time while others are more than usually boisterous. Either sign can lead us astray if we do not know our pigeon. Another point of detail is that rarely will you find a pigeon in good form which has not clean feet which are usually warm, indicating good circulation. Moreover, although you may think that my enthusiasm is running away with me, I declare that often their rings are shining as if they had been polished! Someone is bound to ask what about bloom as an indication of the presence of form? We know very well that some pigeons carry more bloom than others and while we like to see it on a pigeon it is one of those factors which can deceive the unwary. The outward visible indications of the presence of form can be divided roughly into two groups. The first group consists of those which change slowly and can be present without the pigeons concerned being in tiptop form: these are the nesting conditions, the cast and growing flight and bloom. The second group is those which are more delicately balanced and may change completely within a few hours: appetite, the tone of the feathers and, last but not least, the clear bright unwinking eye. The second group is those which give us the up-to-the-minute indications of how a pigeon is. One day they may be all present and then something happens and they are gone. The poor or gluttonous appetite, the loose, puffed-out feathers and the dull, blinking eye, all proclaiming something is amiss; a general listlessness instead of the former sparkle. Bloom is the great deceiver. I have seen many a pigeon really ill and yet it was covered in bloom, and it takes a long time for it to be completely lost because of ill health or injury.

Type of Pigeon

From what has been said, obviously the kind of pigeon we want is one of reasonable physical proportion of which, from its breeding, we would expect to possess homing ability and stamina together with the normal emotions, feelings and desires of racing pigeons. The majority of us are impatient and naturally seek for any short cuts which we think may bring success. There is no doubt that the selection of the original stock to start with is very important. Selection on the basis of visible physical characters is comparatively simple; recognition of the presence of invisible characters is not so easy. Like thousands of others I have searched for signs that would tell me if these unseen characteristics were present and as a consequence have followed many a false trail. The signs are there all right, except that they are rarely where we have been looking or what we have been looking for. The phrenologists, if they are to be believed, by studying the skulls will tell us whether the homing faculty is present. Feathering has been studied from head to tail, and theories exist on the steps between the secondary and primary flights. The length and breadth of the wings have been measured and tabulated, gaps between the end primaries insisted upon. Pigeons with light streaks on the quills of their tail feathers are proclaimed to be constitutionally unsound. Certain forms of keels and vent bones have their adherents, and throats, nasal cleft and tongues have also had the attention of the experts. About the eye, there are theories galore.

Eye Theory

In an earlier chapter we had evidence that pigeons were primarily visual creatures, therefore the eyes, the organs of sight, must play an important part in the everyday life of pigeons. Like many other organs they no doubt have secondary functions and it is probable that the influence of the sun on the changing phases in the life of pigeons throughout the year is transmitted through the eye. It would be a mistake to think that the eyes are more important than the organs and parts of the body, large or small. Each plays its part in the scheme of things, but they are all interdependent on each other. The eyes, like every other organ, are specially constructed to carry out an allotted function, no more and no less. The structure of the eye is the same in all pigeons; they differ only in the colour and pattern of the iris (except in the case of malformation or the effects of disease). The purpose of the coloured iris is to beautify the bird and to keep out extraneous light rays. Despite the simple logic of the above facts, periodically there sweeps throughout the fanciers' world some theory concerning the visible pattern of the iris, which has come to be called Eyesign.

There exist two deep-seated failings in human nature which provide conditions favourable to the resurgence and popularity of the eyesign and other similar theories. The first is that throughout the ages men have continually searched for visible signs and potents which they believe disclosed the presence of something otherwise hidden. The second is that we all love mystery; it fascinates us. Thus, whether we openly admit it or not, it is the combined result of these two human failings which at times bewitch us into believing in what amounts more or less to black magic. Within my own experience I have witnessed eyesign and many other fancies run their course like infectious diseases. Once a fancier is infected, no amount of reasoning will stop the course of the infection until gradually disillusion sets in, as it eventually will. I speak from experience since at sometime or another I have suffered from most of these infections which, perhaps, has given me an immunity to fresh outbreaks. Except those who have written books and articles propounding the eyesign theory, the majority of writers have been non-committal on the subject, which is not very helpful to their readers. I propose, therefore, to give my own experience of it and my conclusions, from which the reader can judge for himself.

Let us first examine in detail the portion of the eye that is visible, in order that we may be clear about what we are looking at. When we look at a pigeon's eye, the portion we see is only that which is exposed between the open eyelids; the remainder of the ball-like organ is sunk into the eye cavity of the skull. In the centre of the portion visible can be seen a black circular dot known as the pupil. The pupil is actually a round opening constructed so that it can control the amount of light rays entering the interior of the eye. In bright light it contracts and in dull light it dilates, thus providing good vision under varying light conditions. The portion of the visible eye immediately surrounding the pupil is known as the iris which normally has superimposed on its surface minute particles of colour pigment. The exception to this rule is found in what are termed 'bull eyes' which are the result of the total absence of colour; what we are looking at is the uncoloured iris. In coloured irises the density of the colour particles is normally greater at the outer edge of the coloured portion of the iris, becoming gradually less dense towards the centre (pupil) or pupilary margin. The result of this is that the colour is of a deeper shade at the outer edge of the circle formed, becoming lighter or of another colour towards the pupilary margin. More often than not these coloured particles do not come right down to the edge of the pupilary margin but stop short of it in irregular patterns. This at the first glance gives the impression that the pupil is anything but round. A closer inspection in a good light will disclose that where there is no pigmentation the uncoloured iris can be seen as a dark shade extending to and forming the pupilary margin and it will be noted that it is completely round and even.

It is on the position and pattern of these unpigmented portions of the iris, together with the several circles and other signs, that the eyesign theories are based and from which the self-appointed experts will appraise the merits of a pigeon as a racer and/or producer. The eyesign theory was given to me as a great secret and I still possess, after all these years, the original translations and accompanying coloured plates illustrating the signs to look for, good, bad and indifferent. These vary little from the present-day published versions, except that a little phrenology was thrown in for extra measure. Of course, I fell for this new cult; was it not just what we all had been looking for, an immediate on-the-spot method of sifting the good from the bad? In due course all my birds had been classified by their eyesign according to the book and plates. Being a canny Scot, however, I did not kill the poor eyesign birds as instructed in the book. For two years all my youngsters were classified by their eyesign and recorded in my notebook. At the end of two years, when the first season's youngsters were two-year-olds and the survivors had all flown 480/550 miles, and the second season's youngsters were yearlings who had flown over 300 miles at least twice, I analyzed the results and was disappointed to find that there were equal proportions of each classifications among the survivors and the lost. I know very well that the eyesign experts will say that not being an expert I could not judge properly. The fact remains, however, that I did classify these birds by their eyesign, and if any particular type of eyesign had meant anything, surely by the end of the test I must have assuredly come to recognize the right ones.

As I have already said, I was very disappointed that this wonderful way of selecting the good ones had failed. Since then there have been two major revivals of eyesign, engulfing fresh victims. Moreover, the fact that successful fanciers of the moment have been infected gives the current revival fresh impetus until it again dies out for the time being. A few successful fanciers have had the moral courage to state publicly and in print that the study of eyesign leads nowhere but many others having been committed have chosen just to sit on the fence. In their enthusiasm the majority of adherents of eyesign tend to overstate their case; none of them, to my knowledge, has attempted to prove it by admissible facts and figures. It is true that they will tell of visiting lofts and picking the winners out by eyesign; it seems strange to me that these winners and their wonderful eyesign are only discovered after the pigeons concerned have proved their worth racing. It is an indisputable fact that the inmates of every racing loft are the survivors of many bred and reared, selected by that final arbiter the basket. So we see that before the experts come on the scene, selection has taken place, and it must not be overlooked, as so often happens, that it is the pigeons which are missing which could produce the evidence that would prove or disprove this or any similar theory. Elsewhere, there are professional experts who for a fee will select the pigeons to keep, arrange suitable matings and consign the worthless to the pot or pie. My reaction to such a situation is, what kind of fanciers must they be who unreservedly have to depend on another to do their selection and culling? Certainly no fancier worth his salt should ever be in such an unenviable position. I shudder to think of the hundreds, perhaps thousands of pigeons discarded and killed as each of these theories sweep through the fancy.

The Kind We Want

If there is anything of which I am certain about racing pigeons, it is that the pigeons I want should be from a family which for generations has flown 500/600 miles repeatedly, lived to be old pigeons and produced their kind to the end. What better sign or proof is there of their homing ability, stamina, vigour, vitality and fertility. They are never sick or sorry, and they lay regularly, sit keenly and rear a pair of healthy lusty youngsters without turning a feather and go to nest again with the same fervour as they did to the first nest of the season. Then take all we ask them to do in their stride. These then are the signs we should look for in our pigeons and see to it that they are possessed by every bird we breed from in the loft. Pigeons which do not measure up to these standards will not be much real use to us in any respect and especially from the angle of breeding.

The Prepotent Pigeon

There is yet another sign of priceless value when you can find it in a pigeon or a pair of pigeons. It is something for which we should always be on the look out, and that is the prepotent pigeon. It is a very poor loft in which at least one of these cannot be found, to the extent of being more so than the other inmates. The prepotent pigeons are those which, because of their particular genetic constitution, are capable of passing that happy combination of inherited characteristics which tend to produce pigeons of above-average merits as racers and producers. Unfortunately, we cannot tell these pigeons from simply looking at or examining them, or for that matter from their own performances as racers. The test of the producer is the kind of progeny it produces. With racing pigeons this often takes years to find out, and frequently when we do so it is too late, as the prepotent pigeon has been lost racing. In selecting a pigeon for stock purposes we can only assess partially its potential as a producer from what we know about its parents, brothers, sisters and other near relatives. The real test, however, is what it produces when mated to another pigeon or pigeons. The simple fact that there must be two pigeons in every mating is, I find naively forgotten. So often the emphasis is on one bird, particularly if it is of one of the popular strains of the day, or the champion racer of the loft. Every year I get the job of listing pigeons for sale and by the time I have the list completed I not only have an insight into the owner's methods but of his pigeons individually and in particular the prepotent pigeons past and present. The simple reason for this is that their ring numbers keep cropping up in the pedigrees. Now and then these prepotent pigeons are in the sale list and to me would be the obvious choice if I were buying. Squeakers excepted, when we buy it is usually for breeding purposes, so why not buy the bird which has already proved itself as a producer and thus reduce the genetic gamble. In the light of these facts it may seem strange, but it is nevertheless true that these manifestly prepotent pigeons are very rarely the top-price birds in the sale, whether it be by auction or private treaty. As I said earlier it takes time to test and find the prepotent pigeon. Thus, they are usually four, five or more years old which, in the opinion of many, detracts from their value. It is true that they may only breed for a few years, but if what they produce are of the right sort they can be worth every penny you paid for the parent.

Longevity and Fertility

Longevity and fertility are pointers to a sound constitution and resistance to disease; furthermore, they are passed from one generation to another. On the other hand, the loss of fertility, decrease in hatchibility and young going wrong before maturity are danger signs which cannot be ignored. Cocks retain their fertility to a greater age than hens which, by reason of their sex, are prone to disorders of their generative organs and experience a higher mortality rate than cocks. However, the old hen which lays regularly and produces normal-sized and shaped eggs without apparent distress, from which hatch vigorous well formed youngsters is well worth breeding from just as long as she does so.

Quick Recovery

It may appear from the foregoing that we have to wait a long time before signs appear giving us pointers to the vigour, vitality and stamina of a pigeon. This is not strictly correct; even very young pigeons can display signs of their possession of these qualities. Recently, while on holiday, I spent a Saturday afternoon with a young fancier waiting on his young birds returning from a race. He got three birds quite close together and then there was a long gap before another turned up, to be followed by two or three more before we went off to the clock station. Clock-checking over, we had our tea and later returned to the loft just in time to see a pigeon arrive. Given a light feed and a drink this bird sat hunched up with its feathers puffed, the picture of misery. Looking at it, my young friend said, 'I am afraid that one is finished racing for the rest of the season.' Turning to him I said, 'We will tell better tomorrow.' I understood his rather hasty judgement was based on his comparing this pigeon with the early arrivals sitting there without a feather turned. What he overlooked was that the late arrival had been more than double the time on the wing than the first birds. So I said to him, 'Tomorrow, if it is worth keeping it will be as bright and alert as any of the others.' The time and manner in which a pigeon recovers from a gruelling trip gives us a good idea of how sound is its constitution and stamina. I could relate very many instances of where pigeons which later proved to be champions gave the first indication of their quality by their quick recovery from a gruelling trip when a youngster or yearling, but I am sure such instances will have been experienced by all. Our thoughts turn to such things when the sun in the north-west is turning red and the day is closing fast. We think of the pigeons we have been waiting on for hours and wonder if any one of them is still plodding on, and if so which, as we turn them over one by one in our minds weighing up the chances of each. Do we go over their pedigrees and strain names? On the contrary, we recall and think of the pigeon or pigeons among them which, under similar conditions, battled on to the edge of darkness, and it is on them at this critical time we pin our faith and hope. At such a time we become realists forgetting in the emotion of the moment of our favourite ideas and fancies and cling desperately to the hard facts. Why not stick to the hard facts every time and all the time?

The Future is With the Survivors

Never keep a pigeon simply to look at but set a standard of performance for your pigeons and shed no tears over those which fail. Your future success rests with the survivors. The surest way of building up a successful team of racing pigeons possessed of homing ability, stamina, vigour, vitality and fertility may appear to be the hard way. It consists of racing them out to the bitter end year after year. Eventually, the survivors, being those most fitted to thrive in the environment you have created, will tend to produce progressively more of their kind. You will thus establish a family peculiarly your own, which will do better for you than any other pigeons you may subsequently introduce.

A Changed Environment

It is often found that pigeons that are successful for one fancier are failures in the hands of others. Very often the birds are blamed; but I have my doubts. There are so many factors involved and, as we have seen, a changed environment brings changes in behaviour. Jack T. Clark and I exchanged perhaps half a dozen youngsters each year. Some did well and others did not, but that was no different from those we bred and kept ourselves. Many times Jack has said to me, 'Just look at that pigeon. It has been sent to me by a chap who bought a few birds from me some years ago and it is bred from these. Would you recognize it as being of my family?' I certainly would not. This was not an isolated case for he often remarked about it. He would say, 'Now the birds you send me down are no different from my own in the loft. All I can think is that these others do not handle their birds properly.' It is a well-known fact that close relatives, no matter how they may have resembled each other originally, tend to differ when brought up in different environments. It all leads to this—the kind of pigeons each of us possess, good or bad, are, strictly speaking, of our own making. First of all from the kind of pigeon we are prepared to keep and breed from. I am quite sure that less than half the birds in the average loft are really worth breeding from. Any one who doubts this assertion need only go back over his own breeding records and the further back he can go the more certain I am that he will come to the same conclusions.

Many From the Few

The general run is to take a pair in the first nest from all the mated pairs in the loft. The result of this is that we have a few from the many, when in point of fact it should be the reverse. We should logically breed many from the few pairs we know to be proven producers, judged only by what they have produced. This is no genetic theory but proved over and over again. I know intimately one very successful fancier with perhaps twenty pairs of pigeons and he will only breed from four to six pairs of them. Furthermore, one, if not his most successful racer, now six or seven years old, has never been bred from. It is not that this pigeon has never been mated, but any youngster he has reared is that of another pair. The kind of pigeons the youngsters we rear turn out to be are to a great extent what we make them by the environment we have created for them. The longer I live among pigeons and pigeon-men the more I become convinced that the difference between the truly successful and the not so successful fancier is not because one has better pigeons by inheritance. Money can buy inheritance; with money a man can buy the best bred birds in existence and there is no doubt about it that thereby he gets a good start. But money cannot buy consistent success. The good consistent environment, in other words the 'Know How' of management, is what makes the difference, so Think Again! The man and his methods of breeding and training are what bring out the best in the inherited characteristics. Pigeon racing is no different from many other competitive sports, the winner is the one who makes the fewest mistakes. Thus, to the beginner and others I say look out for the signs of longevity, stamina, vigorous vitality and fertility in choosing your pigeons. These are the sort that will give you a good start and the rest is up to you and your management.

Strain-names

A friend who has read what I have said in this chapter so far has commented, 'But you have said nothing about strains.' When I asked him what he meant by strains, he rattled off a few of the popular strain-names of the day. Personally, I think far too much attention is paid to strain-names, because when all is said and done whether it be as a producer or racer it is the individual pigeon itself which matters. Study the breeding of any successful fancier's birds, or better still make an honest analysis of your own breeding over a number of years and you will be surprised as just how few birds were really worth breeding from. Another rather sobering thought is the numbers you rear to produce a good one. Moreover, remember the famous fancier is no more immune from the law of averages or the genetic chance than each one of us. Please do not misunderstand me; I know very well that the mere fact that certain fanciers have bred and reared a pigeon is in its way a recommendation, as they are unlikely to breed from worthless pigeons, subject, of course, to vagaries of the two conditions already mentioned. The trouble with strain-names begins when the pigeons have been bred for a generation or two in hands other than those of the fancier whose name they bear. We must never forget that at every generation there is a reshuffle of the inherited factors, with results we have all witnessed. If I may use a simile, you sit down with a companion and there is a bottle of whisky and a jug of water on the table, and you pour some whisky into two glasses and your companion adds some water to each glass. On tasting your glass you declare he has drowned your whisky and he thinks his is too strong. No matter how you juggle with these two glasses of whisky and water you cannot bring them to the taste of you both, and certainly never to pure water or whisky. The only way the desired effect can be obtained is to go back to the bottle of whisky or the jug of water. Inbreeding or the mating of close relatives is not the complete answer, as so many have found to their cost. If you wish to retain the excellent qualities of your champion it is essential that the mate you choose for it, no matter how closely related, must also possess the characters you wish to retain. Thus, every son is not a suitable mate for his dam or daughter for her sire and so on, through the several degrees of relationship. It is surprising the facility so many fanciers have for ignoring the fact that every pigeon has two parents, four grandparents and eight great-grandparents, every one of which has contributed equally in its generation. That great-grandson of your champion is also the great-grandson of seven other pigeons; that is, unless you have been indulging in a spot of inbreeding to your champion. Look out for the signs I have already enumerated for these are more important than a strain-name. Find your prepotent pigeon and treasure it. That this is important is written in capital letters throughout the history of the racing pigeon fancy in Great Britain. The pages of *The Creation of a Strain* by Wing-Commander W. D. Lea Rayner, MBE, that comprehensive and authoritative history of the development of the present-day racing pigeon in Great Britain, teems with examples of the influence of prepotent pigeons, both among the importations of the early pioneers and around which present-day fanciers have built their families. You will find that it is the individual pigeon which counts, and the kind we want are those which by their performance, either as racers or producers, make their own pedigree and that of those which follow. There are many signs, but the trouble is that they are seldom what we are looking for.

I recall getting exasperated with a yearling hen which gave me the impression that she had given up her eggs and mate. I found her flirting with several different cocks and fighting with their hens. The result was that I put her among the young birds during her mate's turn on the eggs. In the evening I put her back to her nest box after removing the cock and she took to the eggs. Later I saw the cock re-enter the nest and push the hen off the eggs; he was promptly put away among the young birds. Next day he was basketed for a race in two days' time, which he won easily in Club and Fed: of course as you can guess I had only coppers on him, and was teased by my clubmates about it. I was very cross with myself for allowing my annoyance with the hen to blind me to the true situation.

6 Opinion Versus Fact

A Doo's Clecking

Many widely-held beliefs, for which there is no substance in fact, have been passed down through the centuries. It is widely believed that the two eggs laid by the hen pigeon always produce a cock and a hen. In fact, a family here in Scotland which consists of a boy and a girl is termed a Doo's Clecking (pigeon's brood). We know, however, that a pair of nestmates are not always of opposite sexes but quite often they are both of the same sex. In theory, the sexes should appear in equal numbers in each generation and if we could record the result in large enough numbers this expectation would, we think, be nearly correct. At the end of each season, Jack T. Clark and I, working with the same family, usually found we had a surplus of hens. I was, therefore, of the opinion that we both produced more hens than cocks from our matings. Imagine my surprise when, on investigating the matter, I found that in actual fact we both produced more cocks than hens. Thinking the matter over I came to the conclusion that as a rule hens mature quicker than cocks and survive the training and racing as young birds better than cocks, especially in a family which was slightly larger than average. This is an example of an honest opinion which, when investigated, was found factually incorrect.

Facts and Figures

Intrigued by the result of my investigation of Jack's and my own pigeons, I asked a few other fanciers with the necessary records over a minimum period of ten years to let me have their data. From these records of 831 nests the 1,662 pigeons involved consisted of 860 cocks and 802 hens, which in round figures give a percentage of fifty-two cocks and forty-eight hens. The figures as to the composition of nests are equally as interesting and completely debunk the old belief that there is always a cock and hen in each nest. Of the 831 nests, 165 contained two cocks, 136 contained two hens and the balance of 530 contained the proverbial cock and hen. When considering facts and figures in connection with the problems of inheritance among pigeons a large number are required to form a judgement. In this connection it should be remembered that it is usually the eggs you were keenest about and what they would produce, which came to grief in some way or another. Moreover, there are the hundreds of eggs laid which are never allowed to hatch; perhaps what we seek was in some of these. I remember the fancier who secretly hoped that one day he would produce a pigeon like the old cock. Last year he gave another fancier a pair of eggs to carry away in his pocket and later told me that from one of these eggs was hatched the exact replica of his old favourite. I tried to console him with the thought that while this pigeon may look like the old cock it may not have his other sterling qualities.

Colour Inheritance

I am afraid I rather damped the delight of another fancier at having bred a Barless Mealy cock from his family, which are predominately blacks but originated many generations ago from a very good Barless Mealy cock. My friend is quite convinced that the arrival of this latest Barless Mealy is proof that the old Barless Mealy is still exerting his influence after all these years. In fact, however, the latest Barless Mealy is the direct result of his introducing a red hen into the family. When he told me about this he had another Barless Mealy in the nest which he hoped would be a hen, though there is nothing more certain than that it will be a cock. Reds from a dark-coloured cock and a red hen are always cocks. Similarly, the dark-coloured youngsters are always hens and any dark-coloured youngsters from a pair of reds are hens. Youngsters from a red cock and a dark hen can be of either colour and sex. Any youngster with short down (the fluffy yellow hair with which newly hatched youngsters are covered) carries the genetic character 'Dilute' which turns Black to Dun, Red to Yellow and Brown to Khaki and Blue to Silver. If neither parent is a Dun, Yellow, Khaki or Silver, such youngsters are always hens. A fallacy exists that short-downed youngsters are no good: pay no attention to this idea. This is useful information to have as you can tell the sex of a youngster the moment all its down or pen feathers proclaim its colour.

Belief in Facts and Figures

That the maiden eggs from yearling hens have special virtues has been given prominence and publicity by well-known and successful fanciers and in my early and more credulous days I believed them. Later on, I took the trouble to sift the facts, and while at Aldershot I collected some more data on the subject from the fanciers who passed through training courses there. A similar belief is that at a first mating, especially if one of the partners to it has been brought in from another loft, the chances of breeding a good bird are increased. Let us take the maiden eggs idea first. The data disclosed that 50 per cent of all matings involved a yearling hen, that this first pair of eggs were the only ones allowed to hatch from these hens in their yearling stage and were often the only ones laid, as a number of the birds were lost later racing. So you see how heavily-loaded the chances are on the side of the maiden eggs, because between 80 and 90 per cent of the young reared from yearling hens are from the first round of eggs. Let us now turn to the case of first matings. From the two censuses taken it was found that 90 per cent of the matings in the lofts investigated were first matings; thus, the chances of a first mating producing a good bird are nine to one. Therefore, if yearling hens and first matings do not produce a high percentage of good pigeons they ought to, since their numbers exceed all others by a wide margin. It has been suggested that the pigeon brought in from another loft is, as the result of the change of environment, more likely to produce a champion. There is no doubt but that the records show that very often the champion has been the result of the introduction of a bird from another loft. However, this is hardly likely to be solely the result of the change of environment. There are several other factors much more likely to be the contributory cause of the favourable result from such matings. In almost every case where a bird has been brought into a loft and bred from, it was a pigeon selected for the special purpose of improving the standard of the loft. I think you will agree that it would be very foolish to bring in a pigeon which was inferior in any respect to your own. It stands to reason, therefore, that any pigeon introduced into a loft for breeding purposes is likely to be mated to one of the best birds in the loft. That being so, if such matings do not produce the good birds, what was the use of it? Moreover, there is a very good reason for this from the genetic point of view: that well-known phenomenon commonly called hybrid vigour or, technically, heterosis. This frequently occurs when two pigeons belonging to distinctly different families, especially inbred families, are mated. The resultant offspring are usually found to be more vigorous than either parent, and the combination may well produce the champion racer. But there is a snag, in that pigeons produced from such matings are more or less useless as producers of birds of the same high quality as racers as themselves. This is a source of great disappointment to their owners. There are two schools of thought on this matter: those who practise inbreeding in its varying degrees of near relationship and those who out-cross all the time.

Pooled Information and the TRPA

In the closing paragraph of *Win With Olympic*, I wrote, 'No single fancier can hope to gather from his own experience the data and details required to solve many of the problems. . . . Some method of pooling and checking information seems a necessity.' When I wrote this I had in mind the Thoroughbred Racing Pigeon Association. Since then the TRPA has carried out some research and an analysis of the results show that of the breeding of the large number of winners investigated, 63 per cent were the result of varying degrees of inbreeding. On the other hand, the few really outstanding champions were the result of an unrelated parents' mating. We have to be careful not to jump to hasty conclusions because 70 per cent of the parents of the successful racers were inbred in some degree, pointing plainly to the fact that the out-crossers were indebted to the inbreeder for their success. Some other interesting information was extracted from this research. The average age of the successful sires was three and a half years, and that of the successful dams three years. Here again we have to be careful, because we are dealing with an average, and further investigation has shown that the average age of the inmates in the majority of racing lofts works out at slightly under three years. A census I took some years ago disclosed that the age distribution in the average loft was 42 per cent yearlings; 24 per cent two year olds; 16 per cent three year olds; 8 per cent four year olds; and the balance older birds. The average age in such a loft worked out at two years, which demonstrates that the number bred from, favours the younger pigeons. It is suggested that, like maiden eggs and first matings, first-nest youngsters are more likely to turn out winners than second-round youngsters, but my data shows that for every second-round youngster reared there are four first-round youngsters. The reader can easily check the accuracy of these figures from his own records, and the greater the number of years involved the truer will be the average result. Despite the fact that the average age of the successful sires was three and a half years, and that of the successful dam three years, only eight sires and eleven dams in every hundred were yearlings. The oldest sire was ten and the oldest dam nine years old, which are facts as opposed to opinion extracted from the records of a great number of pigeons and variety of circumstances than it is possible for any single fancier to obtain personally.

H. J. King's 'Twilight'

I feel sure that the breeding of *Twilight*, bred and raced by H. J. King of Camberley, winner of the National Flying Club Pau race in 1959, will be of interest and give many cause to think again. This bird was bred 1954 NURP54FT 114, a Light Blue Chequer hen. As a yearling she won 1st Weymouth (90 miles), 3rd Guernsey (150 miles), 1st Bordeaux South Coast Combine

(450 miles) and as a two-year-old, 1st Guernsey; 4th Sect., 8th Open N.F.C. Pau (550 miles). At three years old she was a winner from Nantes, then 3rd Sect., 4th Open N.F.C. Pau. At four years old, she won 6th Sect., 59th Open N.F.C. Bordeaux. At five years old, she won 4th Weymouth, then 1st N.F.C. Pau 1959. Her parents were unrelated, but note how they were bred. Her sire, Blue Cock NURP49DM 2423, was bred by F. Thomas of Hawarden, Chester, from Champion 91 when mated to his own grand-daughter, *Fearndale Queen*, winner of N.W. Nantes Club, etc. Her dam, Dark Cheq. Hen, NURP48 BA 1477, was bred by the late R. A. Hamblin, Leicester, from the champion *Newmarket Queen*, winner of 1st and Kings Cup North Road Championship Club and of six first prizes in succession, etc., when mated to her own son. I have in previous chapters stressed the importance of longevity in a family as an indication of its vigour and vitality, *Twilight's* sire was five years old and her dam six when she was hatched, and when eleven and twelve years old, respectively, they fetched £121 as a pair when sold by auction in October 1960: someone knew what they were about. Her sire's sire was fifteen years old and his dam five years old when they bred 2423. *Twilight*'s dam's sire was a yearling and her dam ten years old when they bred 1477. Further proof of the longevity and vigour of these two families from which *Twilight* is descended is that her sire's pedigree goes back twenty years in two generations on one side and twenty years in three generations on the other. On her dam's side the pedigree goes back twenty years in three generations and sixteen years in three generations. The above facts are well worth pondering over, i.e. *Twilight*'s racing performance beginning with 450 miles as a yearling. This is nothing new. S. P. Griffith's famous '459', forty-five years ago, had an equally brilliant performance simply because she continued to be sent racing. Secondly she was the result of the crossing of two inbred and unrelated pigeons each with a record of longevity. I am sure that any investigation the reader cares to make on the subject will reveal ample confirmation, provided the numbers involved are large enough to give a reasonably good average.

Perseverance is Essential

Every pigeon has its day, no matter whether it be a champion or just a good average pigeon. One day the conditions suit it and the best performance of its life is the result. So many pigeons are considered nothing out of the ordinary for the simple reason that they never get the chance on their day and the fault lies with us, their owners. One of two things usually happens. The pigeon is tried out once or twice at 500 miles and if it does not score it is discarded as slow and useless; or, even worse, it scores well from 500 miles and is thereafter kept for stock without there being any proof whatsoever of its ability as a producer. Pigeons which will fly 500 miles are not so plentiful that they can be discarded, nor can we afford to take a good racer off the road for stock. If it is the kind we want to breed from it will fly 500 miles over and over again.

John Taylor's *Newslad*, twice winner of 1st British Section, Barcelona, was sold for £500 in November, 1959 and *Twilight* was sold for £625 in October 1960. There is no doubt that it was the repeated performances of these pigeons which enhanced their value. In the years 1952–3 and 1954 Alex Galloway of Leslie, Fife, sent his Blue Hen SURP-49-E 8612 to compete in the Scottish National from Rennes 563 miles. It won 122 Sect., 320th Open in 1952, but in 1953 and 1954 it was without success. In 1955, at six years old, 8612 was sent back again, and this time she won 6th Open and Section from 6,201 entries sent by 2,485 members. The following year, 1956, she was sent again and this time the seven-year-old hen proved to be the winner in an exciting finish, winning the King's Cup, the Gold Cup and a string of other trophies. The second bird was a three-year-old cock of E. R. Williamson of Kirkcaldy, and the third bird was a six-year-old hen of Robert Strachan of Invergowrie, SURP-50-L 3740, which had been fourth Open and Section the year previously. The point I wish to make is that Mr. Galloway might have given up sending 8612 after two failures, and both Mr. Galloway and Mr. Strachan might have been forgiven stopping these ageing hens after their successes in 1955, in which case these good hens would not have had the chance of proving just how good they were. These pigeons have all the signs I look for in the kind of pigeons I want to keep, namely the ability to fly 500/600 miles over and over again, live to be old pigeons and produce their kind right up to the end. Stamina, fertility, vigour and longevity are inherited, but they must be carefully cultivated by breeding only from the pigeons known to possess them. When inbreeding is mentioned we immediately think in terms of the mating of close relatives. When, however, we inbreed to a characteristic or group of characteristics as those listed above, the partners to the mating need not be closely related, if related at all. A classic example of this is the breeding of that great hen *Lanimer Queen* bred and raced by Willie Hamilton of Lanark. She was in the prize-money in five successive Scottish Nationals from Rennes 526 miles and was timed at her sixth attempt. She brought the highest price at the late Dr. William Anderson's dispersal sale.

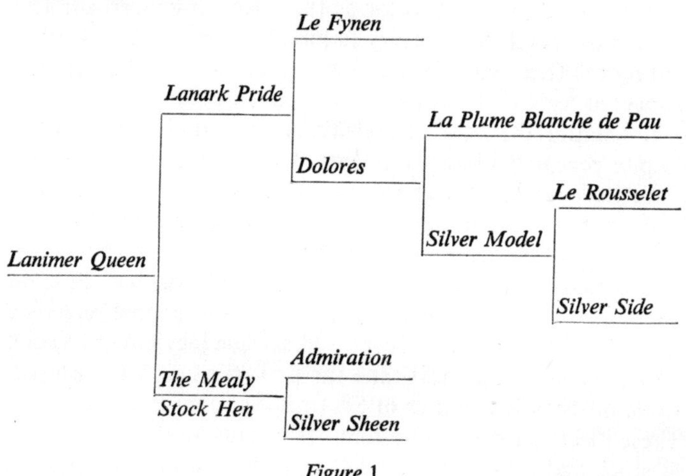

Figure 1

From her pedigree we can see that *Lanimer Queen* was a granddaughter of *Le Fynen* and *Admiration*, a great-grand-daughter of *La Plume Blanche de Pau* and a great-great-grand-daughter of *Le Rousselet*. These four champion cocks were recognized as being the best of their day. *Le Fynen* was bred and raced by M. Tremmery and between 1933 and 1937 his winnings were phenomenal and he was awarded the title of European Champion at the Olympiad of 1938. *Le Fynen* was a picture to look at and handled as we expect a champion should. A point to note is that his sire was a full brother to Tremmery's famous Pau Hen, second in the Belgian National from Pau in 1930 and 1931 beaten each time by *La Plume Blanche de Pau*, another handsome pigeon, which I have handled many times. In honour of his racing achievements he had a gold band on his leg presented by the King of Belgium.

Earlier on I said that every pigeon has its day, but even the best of them have an off day. Pigeon racing is influenced by many factors which will come readily to the mind of the reader. When reading the records of prominent winners their failures are seldom if ever mentioned; perhaps not intentionally but because they do not seem important. There is nothing further from the truth, at least so far as the owner is concerned. The whys and wherefores of every failure demand an investigation and more often than not the fault will be found to lie with the owner: at least that has been my experience.

No matter the field of endeavour, we admire those who, despite repeated disappointments, by perseverance try again and succeed. Close on twenty years ago I listed for sale some birds for the Leishman Bros., of Douglas Water, Lanarkshire. I was intrigued by the success and the source from which it came. The foundation birds were initially a pair of birds bought from the late H. R. Keer of Ipswich in 1943, now denoted reference A and B. Later on at the Keer dispersal sale they bought a cock now reference C. By this time the first pair had set Leishman Bros. off on a long series of Scottish National F.C. successes. These three pigeons bred a number of outstanding performers at stock and on the road. What wonderful prepotent pigeons these three proved to be, living to a ripe old age and reproducing their kind right up to the end. Take Red C: he produced a winner from Dol 500 miles and Rennes 540 when thirteen years old and his daughter, bred in 1948, was the winner of 14th Sect., 28th Open S.N.F.C. Rennes 1951 and in 1952 she was 8th Sect., and 14th Open. She flew Rennes in the next three years and in 1955 she won 11th Sect., and 78th Open. Now you see what I mean by every pigeon having its day. *Holm Queen*, as this hen was named, did not lay in 1959 and 1960 but in 1961 she laid a pair of fertile eggs at thirteen years old. The inmates of the loft today go back several times over to ref. A, B and C and their most recent success is the winning for the second time the Lanarkshire Social Circle 'Gold Cup', 1972. The winning of this coveted trophy set the seal of success on both the man and his pigeons. The programme consists of five races from France, three from 500, one from 550 and one from 600 miles. In each race competitors can nominate four birds, but only a bird nominated counts. These simple conditions can break a competitor's heart. First you can time several birds before you get a Gold Cup nominated bird; failure to time one and you are out of the cup for the season. Pigeons that can win in this kind of competition have all the signs I look for in the kind of pigeons I want to keep, namely the ability to race and win from 500/600 miles; over and over again they live to be old pigeons and produce their kind right up to the end: what better proof of their stamina, fertility and vigour. These characteristics are inherited but they must be carefully cultivated by breeding only from birds known to possess them.

Fresh Angles

The longer I live and observe the more convinced I become of the need to think again from fresh angles about the many practices which from custom and usage have become the accepted thing to do. For instance, the custom of feeding pigeons in transit by the convoyers. Pigeons basketed in the evening for liberation the next day should not be fed by the convoyers. Where they are to be twenty-four hours or more in the basket before liberation they should be given a light feed in the early evening before liberation. A few enlightened organizations have adopted the above practice with gratifying results. If you try to put forward such a proposition to the Annual General Meeting of your club or Federation, however, do not be surprised if its creates an uproar. You will be accused of cruelty and much else besides. Yet I know that to set a pigeon off on a journey long or short with food in its crop is handicapping it just as effectively as the racehorse is handicapped by the weight it has to carry as a penalty for its previous success, or for its age.

So many of the opinions and practices in pigeon racing are based on plausible hypotheses which, unfortunately, so few have taken the trouble to prove or disprove by reference to the actual facts. There are those who believe that a pigeon is handicapped when sent racing with a new flight just bursting from its follicle, and will quote the fractiousness of the teething child in support of their belief. I recall many years ago being shown round the premises of a well-known purveyor of pigeon-feeding and of how, in his sales talk, he pointed out how nicely balanced was a particular mixture of pigeon-feeding. He did not tell me about the protein and carbohydrate contents but of how a pigeon fed with this mixture and sent on a long flight would first of all be sustained by the millet, dari wheat, peas and lastly by the beans in the mixture. He painted a vivid picture of a pigeon on the last stages of its journey digesting this last bean and thus having the strength to carry on. Of course I took it all in and bought the mixture; I admit I swallowed most things I was told and that is why I am so aware of the need we all have to think again. I was told, and believed, that it was wrong to mate two pigeons of the same colour and pattern and if I got light washy colours it was a sign of degeneration. The textbook of those early days rated Mendel's theory as of absolutely no practical use or value whatever to the breeder of racing pigeons. I rate highly many of the observations of those early pioneers in the sport and marvel at times that with their limited knowledge, at least compared with what is available to us today, how near they came to the truth in their interpretation of what they observed. It is in the interpretation of what we observe that I urge the reader to think again, making full use of up-to-date knowledge on the subject. If those early writers had persevered with Mendel's theory they would ultimately have found the answers to many things that they observed but could not explain. The consequences are that the racing pigeon fancy is lagging behind the breeders of other livestock in very many respects. The progress others have made in methods of breeding, the art of feeding, in quantities and values, the prevention of disease and the control of parasites common to the animal or bird in which they are interested, may be taken as an example of what can be done. However, we must not accept their findings as being the same for pigeons, because while there are many likenesses there are just as many differences and the problems concerning racing pigeons must be worked out by ourselves. I realize the work done in other spheres is all a matter of economics. The production of more milk, eggs, beef, mutton, pork, grain, roots, fruit and pure and better strains of bacteria for the production of yeast in making of bread or the brewing of beer are all commercially important and that money attracts brains. But what have the pigeon organizations spent in the way of research into any aspect of the sport? Nothing, as far as I am aware.

Casualties in the Loft

Racing pigeons, fortunately, are a hardy variety of the species and the sick or infirm have no place in the scheme of things. Broadly speaking, casualties in the average pigeon loft are roughly grouped by fanciers into three main groups: injuries, canker and going light. With injuries it is amazing how quickly and well pigeons recover from ghastly looking injuries when helped by a bit of rough surgery in drawing the wound together with a needle and thread. Canker is a term loosely applied to any sore displaying a yellow cheese deposit. The presence of this yellow substance denotes that the wound or sore is septic. Wounds from visible external causes excepted, these septic sores are the visible symptoms of several diseases found in pigeons caused by virus and other infections. In mammals, septic sores discharge liquid pus but mammals lick their sores, keeping them comparatively clean. You can imagine the mess a pigeon would get into with any sore discharging liquid pus. It appears that mammals possess a type of white blood cell which has the power to digest dead bacteria and other foreign bodies. Birds lack this particular white blood cell, hence their solidified pus. The presence of this solidified pus is just a symptom and must be taken in conjunction with the other symptoms in diagnosing the disease. True canker usually attacks the mouth, throat and gullet, particularly with youngsters in the nest, which are best killed. It is not so common in older pigeons kept under reasonably hygienic conditions free from dampness. Going light (emaciation) is a symptom of many pigeon diseases and indicates a loss of flesh. No disease brings about such a rapid loss of flesh than a severe attack of coccidiosis in your birds. Coccidiosis, as is true canker, is caused by a protozoan organism. The two diseases are also similar in that birds which recover from a mild attack of either appear to be immune from further attacks of the same organism. Coccidiosis is a specific disease caused by a microscopic parasite called coccidia and, in general, the parasite affecting one species of animal or bird is harmless to those of another species. It is annually the cause of heavy losses among poultry, consequently, most of what has been written about the disease and its prevention and treatment concerns the findings with poultry. The parasite spends part of its life cycle in the intestines where it attacks and damages the intestinal walls and later is passed out in the infected bird's droppings in large numbers. At this stage it is harmless, but given favourable conditions of moisture and air in a minimum period of forty-eight hours, changes take place and the parasite again becomes infective and if picked up with contaminated food or water, it infects or re-infects the birds concerned. Fortunately, in a high percentage of racing pigeons' lofts the droppings are cleaned out regularly, thus breaking the life cycle of the parasite, or at least reducing the number allowed again to become infective.

Powdered Droppings Method of Loft Sanitation

Paradoxical as it may seem, fanciers who knowledgeably practise the powdered droppings method of loft sanitation are less subject to outbreaks of the disease than the majority who scrape and scrub. The crux of the matter is absolute dryness which which is allied to good ventilation. Even a small area of dampness around a water fountain provides optimum conditions for the coccidia oocysts to sporulate and become infective. It appears that the majority of adult pigeons harbour a few coccidia in their intestines without ill effect: apparently they have built up an immunity and can resist further infections. These birds are potential carriers and, given favourable conditions, the sporulation of the oocysts passed in their droppings infects other pigeons. My personal experience of the disease among racing pigeons is that, provided the loft is thoroughly scraped out (not smeared out) and is bone dry, or the Powdered Droppings Method is intelligently employed (*see* Appendix B), there is not much to fear from coccidiosis. It is true that in the cleanest of lofts the occasional young birds will contract the disease (the symptoms are loss of appetite and rapid loss of flesh). I have found that in the majority of cases the victims have had their resistance lowered from some other cause, e.g. from a night or two out, through lifting from the loft or in returning from a training toss. Often these youngsters on their return will be found picking around on the loft floor and thereby increasing their chances of picking up the infection. The modern conception in the control and treatment of coccidiosis is not to try to keep the birds free from coccidia, because no immunity to the disease can be built up unless the birds ingest the parasite. What has to be avoided is an overdose or repeated infestations following close on one another, particularly with young birds before they have had time to build up an immunity. All the well-known drug manufacturers have a coccidiostat on the market under their own trade name. Any one of these can be used with confidence, provided the maker's instructions are carefully followed, as there is a danger in overdosing. The normal method of administering the drug to pigeons is in their drinking water, and in the case of several birds being infected at the same time, all the birds in the loft can be treated as a preventitive and cure. I have used May & Baker's 'Embazin' (sulphaquinoxaline) with satisfactory results. The treatment takes eight days: three days with the drug in the drinking water, two days without the drug and then three days with it. Young birds being fed by their parents in the nest did not appear to be upset in the least, but I would not care to race birds under treatment or for a week or more after the treatment or for a week or more after the treatment had ceased.

Ventilation

It may be only a coincidence but in my experience, the greatest number of outbreaks of disease among pigeons has occurred when lofts have their greatest number of inmates, i.e. when the final old-bird races are on and young birds have started training. This is particularly true during a period of warm, humid weather, when I suspect inadequate ventilation is at the root of the trouble. The most common troubles are of a respiratory nature and young birds are the chief victims, and overcrowded race panniers a suspected vector.

Time was when I thought I knew all there was to know about ventilation but I have been forced to admit that there is a lot more to it than appears on the surface. Pigeons need air and lots of it and one would think that there cannot be much wrong with an open-fronted loft. I have found that they not only require air, but also air that is continually being changed. There must be movement of some kind throughout the whole loft. Light a cigarette and see how the smoke from it drifts, and if there are places in the loft where there is no perceptible drift your ventilation is bad. I have noted lofts with open dowelled fronts about 1½ to 2 feet in depth running the whole length of the loft, yet the ventilation in them was far from good. On the other hand, lofts which appear to be entirely closed up are much better ventilated. When a pigeon breathes, it takes in air consisting of approximately 80 per cent of nitrogen and 20 per cent oxygen. In the process of respiration there is an exchange of carbon dioxide for oxygen, with the result that the air expelled contains 80 per cent nitrogen, 16 per cent oxygen and 4 per cent carbon dioxide. If, therefore, the air in the loft was not changed, its oxygen content would gradually become depleted, with the result that its inmates would become devitalized and their resistance to disease impaired. It should be realized that a great deal of the vitality and vigour of racing pigeons is acquired by braving the elements. On the other hand we must protect them from rain and dampness and, naturally enough, the more air space per bird the longer the oxygen in the loft will last. Big, roomy lofts are not conducive to easy control of the inmates, besides which so many fanciers are restricted for one reason or another in the size of the loft they can erect. The answer to the problem is not so much in providing air space but in having the air in the loft constantly changed by circulation. It is elementary knowledge that the pigeons themselves warm the air by the heat of their bodies and expelled breath. Warm air rises and this provides a certain amount of circulation of the air within the loft, but it is not enough. Some other means of providing adequate and continuous circulation is required and each loft, because of its design, presents its own particular problems. One can be pardoned for thinking that the open-fronted loft must provide all the air that is required but every loft is full of air; the problem is its free and continuous circulation. If my experience is any criterion, the first step towards reasonably good ventilation is not, as might be expected, in getting air into a loft but in getting it out. The most efficient means of doing this is to fit to the highest point of the roof extraction cowls or louvre-board outlets designed to draw the air out of the loft, no matter the direction of the wind. It is not sufficient to have openings at the highest points of the roof which, with the wind blowing in one direction, are all right, but when it is blowing from another direction it is all wrong. This is an important point and requires careful thought and often some experiment to perfect. Having provided an efficient method of drawing air out of the loft, due consideration must be given to the best means of letting air in. There will be very few who have not experienced the effect of a big roaring fire and through every crevice around the doors and windows of the room cold air is rushing in to replace that drawn up the chimney by the fire. Air inlets should be widely spread near the floor level, and the incoming air should be baffled to disperse it throughout the loft. The design of many lofts, because of traps and other apertures half-way up the side of the loft, prevents the free circulation throughout the loft by short-circuiting it between the floor and the roof, leaving areas of air in the loft undisturbed. I have found that lofts in sheltered places present greater difficulties in providing constant ventilation than lofts in more exposed places. Every loft, because of its design, structure and location presents its own problems of ventilation. In a well-ventilated loft, spilled water dries up quickly, there is a freedom from odour and the droppings dry up and disintegrate into a fine powder which in turn speeds up the disintegration of fresh droppings. The effect of bad ventilation is insidious, the birds look all right but lack verve, and a lot of damage can be done before it is realized that something is wrong. A fancier was losing an excessive number of young birds in training and racing, in particular the older birds. Investigation disclosed that there was a pocket of undisturbed air around the higher row of perches normally occupied by the oldest youngsters. Consequently the birds on these higher perches were debilitated from breathing this foul air. The effect was intensified by the fact that his birds were confined to the loft all day except for an hour of exercise in the evening, as they had been eating a neighbour's lettuce. Additional louvered apertures at floor and roof levels cured the trouble.

7 Errors of Judgement

Fortuitous Affairs

I have already said that it is he who makes the fewest mistakes who wins consistently. It is natural, therefore, to think that we should profit from our early mistakes by seeing that they do not happen again. But do we really? All too often we keep repeating the same mistakes over and over again, simply because we do not know what we are doing wrong. He is a wise man who can profit from the mistakes of others and, peculiarly enough, we can see the mistakes of others more readily than we can our own. Thoughtlessness is at the root of most of our mistakes and troubles and it is about the mistakes I have made and those of others I have witnessed that I propose to write. I believe that no one in all honesty will disagree with me when I say that many wins by our pigeons are fortuitous affairs in which we took no active part. The unpooled or lightly pooled winners are glaring examples of the owners' lack of observation. Pigeon racing is a many-angled affair and we are inclined to become too taken up with one angle to the neglect of the others.

Repeat Performances

I have experienced a measure of success in getting pigeons to repeat a performance from a given race point the year following their first success. I kept a table showing what pigeons were in the first three to the loft from each race point and consulted it regularly. All other things being equal I pooled these pigeons as each race came round. For example, a hen, 6273, as a young bird from the first race won 1st Club, 4th Fed. As a yearling she was 3rd Club and as a two-year-old again 1st Club, 5th Fed. and on both occasions she was sent feeding a small youngster.

As a three-year-old she was my second bird winning 6th Club, but this time she was sent unmated. A good record from the same race point, I think you will agree. Subsequently, in her fourth season racing she won pools from 200 miles and was my first bird from 450 miles; on each occasion sent feeding a small youngster. So you see from this that the nesting condition was equally as important as the race point. Another example is that of a yearling hen which won 1st Club 110 miles and 2nd Club 140 miles: she was sent hatching to the 110-mile race and feeding a week-old youngster to the 140-mile race. About a month later, sent hatching, she won 2nd Club from 300 miles, velocity 1,352 and a week later with a small youngster 3rd Club velocity 1,100. At the end of the following season I suddenly realized that this hen had done nothing of note. Looking over my records I found that she had lost her mate early on and had gone to the 110-mile race unmated, and to the 140-mile race just mated. She did not go to the 300-mile race as she had just laid, but was sent to the 350-mile race sitting eight days on eggs. As she had scored in this race the previous year I pooled her but she was only my third bird and too late. My first bird was a cock and this was the third successive race in which he had been my first bird. His nesting condition in these three races was sitting four days, sitting eleven days and, for the third race, sitting seventeen days; I slipped a small youngster under him and up he came to win 4th Club, velocity 949. The mistake I made was that I should have kept the hen for a race two weeks later when she would have had a week-old youngster. With the cock I thought it unlikely for him to come up three weeks in succession. However, if I had given the matter more serious thought I would have realized that instead of going off form he was actually improving.

Winning Combinations

Pigeons will often score two weeks in succession if caught right. The likely combinations are: sent hatching the first week and, if they respond, sent with a small youngster the next. Another combination is to send a cock driving one week and again the next week sitting a few days on eggs. Sometimes it is possible to slip a youngster for an hour or so under a bird sitting over fifteen days on eggs. If it responds, try to get it to take to the eggs and repeat the manoeuvre the following week, and the next week race it to a small youngster. When attempting such a manoeuvre make sure the birds are sitting keenly. There is no surer way of spoiling a good pigeon than to engage it in racing when it has gone stale sitting on eggs; it pays to wait until the next mating cycle has got under way. Some may be inclined to think that when a pigeon has come near the top sent sitting a few days on eggs, the next time they will try it sitting a bit longer in the hope of improving its performance. More often than not a repeat of the same condition is the safest course.

Lessons from Experience

One season I had two late-bred cocks, nest-mates bred from a son of a Scottish National winner. When I mated my other birds up I put these two cocks in an empty compartment and gave them a two-compartment egg box, nailed to the wall, to perch in. They fought and chased one another all day long and when put on the road one or the other was my first or second bird in the first three races. Immediately I had a spare hen I mated them up thinking it would improve their performance. That was a blunder which I have never repeated: the lesson is to leave well alone. I recollect Jack Clark ruefully telling me of how he was misled into making a blunder. He had prepared two cocks, a Red and a Chequer, to send to the Manchester Flying Club Open Marennes race. Both had good racing records and he was watching them closely in order to decide which to enter in the big £5 pool. He had made up his mind to pool the Red cock, but when he entered the loft to basket them the Chequer cock was buzzing around, cracking up and down from his nest and perch to the floor, full of the joy of life. The Red cock, however, was sitting quietly on his nest. When he basketed the two birds the Chequer started chasing the Red around and the Red tried to get out of the basket. In the end, Jack put a division in the basket and put the £5 on the Chequer. The result can be guessed: he timed the Red first, in time to win the £5 pool with the Chequer arriving later and too late.

We can all be wise after the event. The Chequer was fit and full of life but at that time of day he should have been sitting the same as the Red. I have noted that some birds when fit become quiet but must admit that I have been cheated by a noisy boisterous cock.

A young fancier had a spare hen which he kept among the young birds, out of the way. The hen took up with a precocious young cock and laid but the egg was smashed. She was given a nest bowl and laid and sat on the second egg. The cock sat spasmodically and little attention was paid to them. However, when basketing the cock for a race, to his dismay, the young man found the egg had hatched and he promptly killed the youngster. The next day he timed the young cock and it was only later, when he found that he had topped the Federation, that he realized his mistake in killing the youngster. The cock was sent to the next race with high hopes; it was a very relieved young fancier who saw his Federation winner turn up very late, evidently upset by the sudden change.

The Reasons Why

I repeat that pigeon racing is a strenuous sport in which we and our pigeons take many hard knocks, and if we are to succeed we have to put up with them and try again. We are inclined to say, in the depth of some disappointment, that it is all a matter of luck and that there is neither rhyme nor reason for what happens. Similarly, we find reasons to blame the race point, the weather, the convoyer, our pigeons, in fact, everything and anything but ourselves. Such an attitude is an admission of defeat; I know, having suffered from it all too often. I recollect vividly two incidents which at the time had me completely beaten. It was the last old-bird race of the season with a handsome special prize to the winner. Judging by the time of liberation and the wind I expected birds to be arriving around 5 p.m. Shortly after five my friend and nearest club-mate came round to tell me he had timed at 4.30 p.m. Congratulating him I asked if it was the white flighted cock, which was by far the best pigeon he had. He said, 'No, it is the Red yearling.' This bird, from 110 miles, had been away for two weeks returning flown down. Here it was a month later, the winner of a race from 365 miles. The second incident occurred the following year. This time the fancier concerned timed the only bird on the day from 250 miles. The winning pigeon had turned up a week before it was basketed after being away a week, returning with a note on its leg saying it had been liberated about 130 miles away. This distance it accomplished in very good time. The reader may think that there is nothing extraordinary about these two incidents, but I admit they troubled me sorely at the time. I knew both these fanciers intimately. They were easy-going chaps and their pigeons sat about the roof-tops when let out for exercise, or when returning from a race, and would not come in when called. They were always scrambling to get a few training tosses at the last minute before racing commenced. Their losses in training and the early races were shocking. I remember very well that when I met them they would invariably tell of the return of some bird lost weeks before and be quite happy about it. If they had a bird they liked it was either not raced or stopped after a few races, only to be lost the next season. Grand fellows and very good friends; pigeon lovers and their like are the backbone of every club. They pay their fees, take their medicine and usually have little to say. While I rejoiced in their wins, as a stickler for the orthodox maxims, the best food, clean water, regular exercise, methodical training and regular use of scraper, I readily admit it was beyond my comprehension how they did it since neither of them was very particular about what was fed or about cleaning their lofts. Today, however, I know why these two birds of my old friends won as I have related, for subsequently I saw the same thing happen over and over again. Because of the environment in which they lived, through overfeeding and lack of exercise, these two birds were fat and lazy and totally unfit to race well. They received a gruelling and returned with all their suplus fat flown off them, but at the same time they were constitutionally sound. In the case of the Red cock, he had a month's rest and was sent feeding a three-day-old youngster, the first he had ever reared. This was his incentive and I am sure he was fitter when basketed than he had ever been before or since. The other pigeon, also a yearling, had certainly flown himself fit for the week he was away. He took up with a hen on his return and was sent driving. Neither of these birds ever did a thing again. The manner and time in which a pigeon recovers from a gruelling gives us a good idea of their soundness of constitution.

My Nantes Hen

As a yearling she spent a week away from a race of 170 miles, returning badly flown down on the 27 May. When she had recovered and moulted a flight she got two 50-mile tosses and remated. She was sent feeding a small youngster on the 6 July to the Scottish Midland Nomination Open race from Weymouth 361 miles. She won this race and flew Granville 489 miles a week later. Visiting the loft of a friend during the past season, I got my eye on a likely-looking yearling Mealy cock. On examination, I found his first primary flight fretted and that he was holding his second. I asked what had happened to this bird and was informed that some two weeks previous to my seeing it he had been missing from a race for two weeks and returned in a very flown down condition. I said to the owner, 'If this cock casts this next flight before the next race send him and if he has got anything in him he will do the distance.' My friend was not very sure about this, but he did send the Mealy, which to his surprise and delight won the race.

Bad Management

The 3402 was a well-bred pigeon but unimpressive to look at; his sire was a son of a brother to that sensational winner *Lakeland Queen* and his dam a sister to *Tutimes*. In fact, he was a full brother to the *Bordeaux Cock* reference 0 at the Clark dispersal sale, the progeny of which fetched the highest prices at the sale. As a youngster, 3402 was twice first bird to the loft, but in the other races he was nearer last. As a yearling, he just kept coming up until the 300-mile race from which he failed to return. In January of the following year he was reported in a loft twenty miles north-east of his own and was returned by rail. Put on the road again he made a fine start by winning 6th Club in the first race sent driving. In the next four he was nowhere, sent to the 5th race 250 miles driving, up he came to top the Federation. He was still driving when sent to the next race. Elated at his success from 250 miles it was overlooked that he had failed from this race point the previous year and fail he did again. The following spring he returned on his own and on the road again his performance was as mediocre as ever but this time the 300-mile race point was avoided and he finished the race programme to 500 miles. There is no doubt that he was mismanaged as he was sent to the 250-mile race sitting overdue on eggs, and most probably stale. If his eggs had been taken away the week before he would have been driving in the condition in which, from the same race point, he topped the Federation. Carefully-kept notes were available, otherwise the above details could not have been recorded here, but what is the good of keeping notes if full use is not made of them?

Full Crops

I an earlier chapter, I mentioned my habit of asking lots of questions at the marking stations. On one occasion one of the cleverest fanciers I know was telling me that he had to substitute two pigeons because he found that two he had intended sending had food in their cops that morning when he caught them for a final check over before basketing them. Going on, he admitted that they may have been all right and that there may have been a reasonable explanation for the food being in their crops from the previous day, but there was no time to find that out. So, rather than risk two good pigeons, not to mention the pool money he had on them, he kept them at home and sent two other pigeons. As it turned out he timed the two substitutes, one well in the money out of the four sent. It should not be thought that these substitutes were, as so often happens, just two pigeons that could be sent. Rather the reverse is true, they had been prepared for this race and up until that morning had been reserves for just such a happening. Some may ask why the birds were discarded simply because they had food in their crop from the previous day. The answer is that rationally-fed pigeons should have an empty crop first thing in the morning.

How They Feel

Just after he had won a 500-mile open race this same fancier was asked if he would send this bird back to some of the longer races later that season. I can still see the puzzled expression on the questioner's face when my friend answered, 'Yes, if it keeps its appetite and casts its next flight!' Another man expressed surprise that this good winner had been sent feeding a youngster, commenting that they do not look good when feeding. Again the reply came, 'It's not how they look, it's how they feel.' Incidentally, this pigeon some five weeks later was a winner from 700 miles. Recalling the above incidents brings to mind another incident while in the company of the same fancier. We were joined by two others whom I did not know. At that moment another arrived, rather breathless, carrying a basket with two birds in it. His friends asked where his third bird was and he explained he had to leave it behind as when he caught it to basket it he found it had dropped two flights in each wing. While his friends commiserated with him on his bad luck my friend, catching my eye, gave me a wink and smiled. Surely here was a pigeon which had turned the corner, giving evidence of being fitter than it most certainly had been for a week or two prior to casting the two flights. There is no better sign of good health and possible form than a good appetite (not a greedy one, which is a sign that there is something wrong) and an empty crop in the morning. Note that a pigeon which is flown down, sick or injured, will continue to grow any feathers thrown prior to being flown down, sick or injured, but will not commence throwing any more until it has recovered. It is evident, therefore, that the casting of a flight is a sign that all is well. As far as I am concerned the pigeon with a full wing is suspect just as much as the one with food in its crop in the morning. It is true there may be some reasonable explanation for either condition but until I know why such pigeons are out of the reckoning. Once a pigeon has has cast its first primary flight, in the normal course of events it never has a full wing again until it has grown its tenth primary.

Widowers

I make no claims to any great experience of racing pigeons on the 'Widowhood System', having only played with it with a few pigeons. What I have discovered is that only a few pigeons respond to this method with anything like consistent success which, when all is said and done, is no different from pigeons flown on the natural system. The star performers will hold their peak form for two or three weeks and then go off. Much has been written on this subject, and many subterfuges have been used to try to overcome this difficulty. The situation, as I see it, is that cocks put on to the 'Widowhood System' become stimulated by glandular activities which bring on peak form. In addition to this, by the method of training and management they are left in no doubt where their pressing physical desires and emotions can be satisfied, i.e. in their nest box. The natural sequence of the mating cycle, calling to nest, broodiness or the desire to sit, follow one another. It is true they can be delayed or prolonged, but in the end the one merges into the other and eventually the bird loses peak form. It will be noted that this occurs at the time when the next primary feather is overdue to be cast. This is the danger period and the obvious thing to do is to get these birds down on eggs until they cast their next flight. Having cast their flight, the eggs can be taken away and a fresh sequence of the mating cycle commenced, with a rejuvenated pigeon for the next period of racing to widowhood.

Tests of Fitness

Irrespective of whether we manage our pigeons on the natural or the widowhood system we will always have the good, average, and bad pigeons with us. Every one of us is responsible for the kind of pigeon we keep. Pigeon racing, I have said, is a strenuous sport; it is also a ruthless one in which only the fittest survive. Every stage of a pigeon's life consists of a series of tests of fitness before passing on to the next. The more thorough our training preparation the better the chance each pigeon has of passing each stage and the less compunction we need over the failures. There can only be one first prize-winner in every race and it is presumed that every competitor is endeavouring to win first prize. At every basketing night for a race the next day, listening to the conversation among fanciers present, it is apparent that numbers count and that the losers of the previous race find consolation in the fact that all their birds had returned. This attitude of mind was one of the great bugbears of my Army experience. Not once or twice but many times a loft corporal and his men would be sent off with a loft of birds with instructions to get them broken to the new site and trained out in a given direction. A few weeks later I would pay them a visit in order to learn how they were getting along. My first question would be, 'How far have you trained these birds out?' That usually started the fun, and here is a typical example:

N.C.O. 'Oh, I have all the birds homed to the loft, sir, and only lost two.'

Me. 'Yes, but how far have you trained them out?'

N.C.O. 'Ten miles, sir.'

Me. 'You know very well that your instructions were to have them out to "x" by such and such a date.'

N.C.O. 'But, sir, I was frightened I might lose too many by rushing them.'

On closer investigation I found that only ten pigeons out of the forty in the loft had been even once to ten miles. In exasperation I had the lot basketed and that forenoon took them fifteen miles on my way to the next loft. Here, the position was in reverse. The N.C.O. had his birds trained out as instructed, but was obviously fearful of what I might say about the birds he had lost in the process. I soon put his fears to rest by telling him I was not interested in the birds he had lost; what was of importance to us all was the birds he had left ready for the job in hand. None of us likes losing pigeons, but if we never lost any we would be in a mess and when all is said each race is a test for the future, and the future rests with the survivors. Think again and you will find that most of our disappointment over lost pigeons has its roots in sentiment, favouritism and wishful thinking, and has little affinity with fact.

Hard Facts

When we get down to hard facts and analyse the results of our matings over a period we find the majority of them, to say the least, are disappointing. That this is so is confirmed by the fact that it is the exception rather than the general rule to find a pair of pigeons mated together for the second consecutive season. We keep changing our matings around in the hope that in one combination the genetic gamble will come off. However, in actual fact what we are doing is to increase the odds against us. Every mated pair of sound healthy pigeons has the potential of producing a good one, but so few pairs are mated together long enough to produce the numbers necessary to give the law of averages a chance. Just count the number you have in the loft from your best pair or for that matter how many you have from any pair, and you will be surprised. During the winter months I always gave a lot of thought to what birds to mate together in the coming breeding season and put the probable matings down on paper. Before making my final decision I would put them together and weigh up their suitability from the purely physical standpoint and try to visualize the type of young they were likely to produce. If the young came up to my expectations in physique and, under reasonable conditions of training and racing survived the young bird racing, I would keep that pair together for another season, provided they also survived the season's racing. If on the other hand they did not appear to breed true to the physical type expected then I would not mate them together again. When I was tempted to try another mating involving a pair which were breeding reasonably good young, I made a habit of mating the cock to another hen by the method detailed in Appendix A, and by this means I did not disturb the original mating.

8 Second Thoughts

The intention of this chapter is to tidy up a few odds and ends for the benefit of both old hands and beginners. I have been influenced in the selection of the subject-matter dealt with herein by the letters received from readers over the last ten years since this book was first published.

Domesticated pigeons, the progenitors of which are universally believed to be the Blue Rock Dove, are a hardy species which has survived for thousands of years, for all we know maybe millions, on the food provided by Nature.

This fact makes me wonder if all the supplements so widely advertised are an absolute necessity. I ask and ponder this question because the variety of foodstuffs pigeons find palatable and will eat contain a great assortment of the so-called trace minerals which pigeons require more than, say, do poultry to keep the bodily processes functioning.

Minerals and Vitamins

Vitamins are similarly situated in that they are found in the majority of seeds, grains and green-stuff eaten by pigeons. I have never forgotten how a lecturer in a biology class summed up the situation. He said that one only has to mention that something is good for a particular pet, be it cat, dog, pigeon, etc., and the owner proceeds to overdo it. The gist of what he said was: 'You do not require to keep a bonfire burning to light your cigarette—all that is necessary is a humble match or lighter.' This brings to mind a familiar sight in east and central Africa: a column of Negro porters on safari with their loads on their heads and one of each group will be found carrying in his hand a smouldering piece of wood, kept smouldering by his swinging arm, ready to be blown into flame to light their camp fire at the end of the day. A final word about vitamins. There are two absent or more widely dispersed in the variety of food eaten by pigeons, but it is known that pigeons are capable of manufacturing these from other elements contained in their diet.

Taking the above facts into consideration I do not think the beginner need worry much about the danger of diet deficiencies, always provided his pigeons are rationally fed (*see* Chapter 3). There is another aspect one must keep in mind and it is this: it may happen that because of the environment you have created (*see* Chapter 2) there are periods during which pigeons desire certain unusual ingredients in their diet of which your method of management deprives them. Take, for instance, their inherent habit of fielding when feeding young: at this time salt is an essential. It should of course be available in some form or another at all times. Salt is one of the ingedients in Kilpatrick's minerals and in health grit (Liverine); proof of this is the fact that both attract moisture and are eaten avidly when feeding young. If these are available I think you will find it reduces your birds' urge to go fielding. Fielding has its advantages and dangers. Nowadays, the farmer uses a variety of pesticides and fertilizers, so you cannot be sure of their safety. It is not uncommon for a bird or two to miss being basketed for a race through being away fielding.

The Mating Cycle

While the sequence of events which follows the mating of a pair of pigeons is gone through very thoroughly in Appendix A, perhaps a brief résumé will not be out of place here. Once a pair have taken to each other, old hens can be expected to lay the first egg around the seventh day; yearling hens may take up to ten days. The first egg is normally laid around 5 p.m. and is just stood over. It is not until 2 p.m. on the second day, i.e. some 21 hours later, that the second egg is laid and serious sitting begins. Normally both eggs can be expected to chip and hatch between the eighteenth and nineteenth day and both eggs will hatch together. This is because the first egg, having only been stood over for some hours, is slow in starting incubation while the second egg starts right away, making up for lost time. In a few days the eggs, if fertile, will have darkened in colour, but there is no need to handle them to see this. About the tenth day of sitting the walls of the crop commence to thicken and by the time of hatching the cheesey substance begins to peel off. This is the soft food which both parents feed to their young by regurgitation. While the crops of the parents contain the hard feeding they have eaten, the soft food, being light, is more easily regurgitated and swallowed by the youngster. Fanciers ever anxious that their pigeons may not lack for nutriment, will feed small seeds and other concoctions with the idea of assisting Nature. What they are actually doing is hindering the natural processes. If these extras do indeed reach the youngsters all they have succeeded in doing is to dilute the soft food. There is no substitute for soft food. It is 100 per cent digestible. The fact that this is so, and that the small youngsters do not appear during the first few days to pass any droppings, has led fanciers to think that the parents have eaten them. The newly-hatched squab is a helpless creature; waggling its head is the only movement of which it is capable—an action which stimulates the parents with the desire to feed it. We need not worry unduly, as one of the final acts, prior to hatching, is that what remains of the yolk sac is drawn into the abdomen which will sustain it with nutriment for the first twenty-four hours, thus bridging any gap that may exist between hatching and the commencement of the supply of soft regurgitated food.

Methods of Keeping Birds Separate

It pays dividends to keep cocks in the same nest box year after year since this prevents needless fighting. Always close the nest boxes of pairs that are away racing. Also close the nest box where a hen has been left to sit on the eggs, which would be between the hours of 5 p.m. and 9 a.m. next day.

With cocks take them away from the next box around 5 p.m. and return them to their nest, with a warm nest bowl, at 9 a.m. Treat hens in the same way, removing them at 9 a.m. and returning them to their nest box, with a warm nest bowl, at 5 p.m.

This can go on successfully for several days. If and when any bird appears restless try putting a small youngster under them, but do not remove the eggs. When the mate returns remove the small youngster, and usually they will go on incubating the eggs.

Ringing

Always check if the ring is still on a newly-rung youngster for a few days after ringing.

Structural Alterations

If you must make structural alterations do it in January. At this time of the year pigeons are least easily upset. The best way to settle pigeons to a change of environment is to allow them a period of free exit and entry.

Newcomers

To break a new hen, mate her to an old cock (aged); he will keep calling her to nest and not harass her with driving too keenly; then, when let out together, he will keep alighting on the loft, thus encouraging her to do so.

Once you have a newcomer homed to your loft there is no better way of improving its desire and facility to return to your loft than to take every opportunity of giving it tosses up to twenty miles with your own birds. Not only does it intensify its fixation to return to your loft but it keeps such pigeons (including stock birds) physically and mentally fit.

Strays

Catch a stray immediately it enters your loft, place it in a basket and take it away from the sight and sound, if possible, of your pigeons. Provide it with water and a very light pick of food. The next day, if it is in fair condition, toss it at some few miles away, taking care that if it should return to your loft it cannot get in. Give a thought as to why a stranger enters your loft. It most certainly is tired, hungry and thirsty, therefore it seeks to satisfy these overwhelming desires; it wants the company of other pigeons and a safe place to perch for the night in your loft. Deprived of these it is less likely to return to your loft when tossed some few miles from it.

When a bird which you have sold or presented to another fancier turns up at your loft, treat it as you would a stray by putting it into a basket, but do not give it food or water. When your own birds are all in the loft after exercising, close your traps, liberate the returned bird and chase it if it attempts to alight on your loft. If you reckon it hasn't time to return to its new abode keep it in the basket overnight and then, after exercising your birds in the morning and when they are all back in your loft, close the traps and chase the bird which, not having received food or water, will, nine times out of ten, clear off to its new loft. Here are a few possible reasons. If it is in the early spring and it is not yet mated up but is possessed with the desire to mate, its memory of your loft and its old mate induces it to return. This will also happen when a bird has just been separated in the autumn, as it will during the breeding season when it may desire salt and other items of diet found in your loft that are absent in its new loft. I have had birds that arrived almost daily, stay around for an hour or two and then clear off; as they never entered my loft I surmised they came for something they found in the fields, lawn and tennis court. The strangest case was that of a hen which the new owner declared was barren but some time later we discovered that when she normally was due to lay she flew the 150 miles and laid her eggs in my loft and then flew back again. It is sound common sense to be guided by the general rules concering the behaviour of pigeons as a whole and to recognize the exception when it comes along and treat it as such.

Accidents

Quite a few both annoying and at times distressing things happen to our pigeons which, I am afraid, are dismissed as just one of those unfortunate accidents. There is, for instance, the hen that gets scalped when penned in with a cock; or the venturesome squeaker which falls on to the floor of the loft and also gets scalped; the odd one that returns late, evidently after being scalped in the race basket; and we can only guess at the numbers which are so severely injured that they fail to return. This reminds me that some of our early losses with youngsters and yearlings are brought about by the victims suffering from travel sickness, which is not helped by birds having been basketed with overloaded crops. I was visiting a loft and suddenly the owner turned to me and said: 'Perhaps you, who delight in knowing the reason why, can answer me. Here is a very good cock, well bred, yet I cannot get a youngster from him because when they are around fifteen days old I come into the loft and find he has pecked them to death.' Well, to tell the truth this had me stumped. Wendell M. Levi, the author of that masterpiece *The Pigeon*, could not tell me anything about this savage pecking by pigeons. He countered by asking me if I could honestly say I had found any pigeon fast asleep, and I had to admit I had not nor have I found any perceptive fancier who has.

The years passed without my finding the reason. Then it happened. I was underneath the loft sawing logs for the fire when suddenly there was a commotion in the loft. I found a pigeon on the floor with practically every pigeon around it pecking savagely at it. When I entered the compartment most of them stood back while others continued to attack this obviously injured bird. It could not get up, all it could do was to flutter its wings in an endeavour to rise. I got hold of it and found it was caught by a toe in a crack in the floor. When I got it free it flew up to its perch; then I saw that it was bleeding freely. I examined its foot and saw that one of its claws had been wrenched off; it was *The Laird*. Thinking this incident over I thought to myself: why hadn't I seen and understood this before. The answer is that previously I had not been conditioned from my experience to see and understand.

What stimulates and provokes pecking? Simply 'the feeble fluttering of wings'. You will have seen its effect hundreds of times in the past and now you can observe its effect in the future by paying attention to a mated pair going through the actions of courtship during which the other birds appear to pay no attention until the hen crouches down and the cock mounts on her back; and as he tries to maintain his balance in the final act of copulation he flutters his wings, and is immediately attacked and bowled over. Our gentle pigeon or dove is looked upon as an emblem of peace; but a different picture is presented by the species treatment of the sick and injured. Who knows but this casting out of the sick and the injured is one of the inherent traits which has helped them to survive as has the fact that they can thrive on a wide variety of diets in climatic conditions.

Nest Boxes

Now a word or two about nest boxes. By a series of tests it has been found that pigeons are more sensitive to the position of an object than to its size, colour or shape in that order. In other words a pigeon will notice more readily the alteration in the position of its nest box, than it will any alteration in its size, colour or shape. Many fanciers paint the fronts of their nest boxes different colours in order that their pigeons can more readily recognize their own nest box. In actual practice and for the reasons given, the most effective way is to have the entrances in neighbouring next boxes in a different position from each other. In conjunction with this, or as an alternative, the internal layout of each nest box may be different from its neighbour on either side or above or below To those who would like to do a little experiment of their own here is a simple test. Say you have a pair of pigeons in a nest box in which the nest pan is in the far left-hand corner; just shift the nest pan to the far right-hand corner and see what happens. The pigeons will notice it at once—in fact they may even refuse to sit on the eggs. If you alter the size of the nest pan, provided the difference is not too exaggerated, they may notice it but will usually take to their eggs all right. And so in an ever-lessening degree do they notice any change in colour or shape. These are useful things to know and to take heed of.

If you desire to be very sure of the parentage of your youngsters it is essential that no other cock should be able to get near a hen between the fourth day after mating and for just over twenty-four hours after she has laid her first egg. By a series of tests I have found that it is the last cock to tread a hen prior to the ovum dropping into the oviduct which usually fertilizes the egg. As it happens normally with the first egg between the fourth and fifth day after they have taken to each other and for the second some 44 to 46 hours later, it is important that no other cock should get the chance to tread the hen but her selected mate. If my readers will watch carefully they will see that a hen at the period when the ovum is due to drop will sit down to the cock without any of the preliminary lovemaking they go through prior to treading. It is interesting and of some significance that when treaded at this period you will notice the hen's wings give a tremor not present at other times. From my knowledge of the inheritance of colour and other markings, the number of young the parentage of which must be suspect is much greater than is generally imagined.

9 Inheritance

What pigeons we will mate together is a subject never far from our thoughts because on the success of our matings depends our future success in racing. We must admit that most of our best-laid schemes in mating have been disappointing and have been brought about, in the main, by the freakish pranks of the genetic chance.

The breeders of other livestock have made great progress by studying genetics for the improvement of their stock, whether in the production of eggs, milk, beef, mutton, pork, fruit or vegetables. Racing pigeon fanciers are gradually paying more attention to this subject, but from conversations and from what appears in print it is obvious to me that much of the knowledge they have obtained has come from articles concerning mammals, including man.

There are several differences between the genetic constitution of mammals and that of pigeons; these give rise to errors, plus the misconceptions of the true situation by fanciers themselves. Genetics has its own specific language, and I am well aware that the technical terms present difficulties into which I have all too often fallen by a misuse of terms, simply because I have not fully understood all they imply. Let me warn the reader that there are no short cuts. One must start at the beginning and progress step by step, for there is nothing to be gained by attempting to pass from one stage to the next before you have mastered the previous one.

Twenty years ago I wrote my first article on genetics and in the intervening years I have tried to create an interest by writing in simple language, avoiding technical terms as much as possible. Unfortunately, without a thorough grasp of the specific meaning and use of technical terms the reader cannot be expected to read and understand books or articles on the subject.

I will endeavour to explain briefly the fundamental laws of inheritance, using the genetic terms and giving their meaning and specific use, together with common errors concerning them where necessary.

To illustrate how genetic characters are passed from one generation to the next, I will use the visible characters found in racing pigeons, which allow the reader to observe them in operation among his own pigeons and those of his friends.

Every living thing consists of a substance called *protoplasm* which is composed of microscopic units called *cells*, which could be likened to tiny eggs. Enclosed in the cell membrane is a substance called *cytoplasm*. In the centre is found the *nucleus* in which are structures called *chromosomes*, which carry the hereditary factors or *genes*. The chromosomes are usually found in matched pairs, each pair being referred to as *homologous chromosomes*; the members of the same chromosome pair are derived one from each parent.

A pair of homologous chromosomes could be likened to identical string of beads lying side by side. Each bead represents a gene and each gene is matched by a similar gene on the other chromosome. These paired genes, which occupy the same relative position (*loci*) on homologous chromosomes, are called *allelomorphs*, and it is the allelomorphs which control the character or different forms of the same character.

If the reader will now place his two index fingers side by side they will give him a fairly good idea of a pair of homologous chromosomes. Just see how the markings on one finger are matched by similar markings on the other. Now assume that the markings are the genes and the matched pair represents the allelomorph.

Take the characteristic colour: it is found to consist of three basic forms of pigmentation, Red, Black and Brown (Chocolate). The genes determining each of these three forms of colour occupy the same allelomorph, and for that reason they are termed *alleles*. When more than two genes are found to be alleles, thus forming an *allelomorphic* series, they are known as *multiple alleles*. Keep this in mind, because while there can be quite a number of genes determining different forms of the same characteristic, no more than two of them can be present simultaneously in the allelomorph of a normal *diploid cell* (a cell containing pairs of chromosomes). This means that we can find in the paired allelomorphs controlling colour, either two red determining, or one red and one black determining; two black determining, two brown determining or one black and one brown determining. When both genes determine the same form of a characteristic all is plain sailing. Two red determining genes will produce an individual of the red pigmentation group, and for that reason it is said to be *homozygous* for the red form of pigmentation. Should, however, the pair of genes comprising the allelomorph differ from each other in the form of the character they determine, one form will express itself and the other form will be suppressed. The form of the character which is expressed and which we can see is said to be *dominant* and that which is suppressed and hidden is *recessive*.

Care must be taken about the use of these two words 'dominant' and 'recessive': they are used only to express the relationship between alleles: also note that it is the form of the characteristic that is dominant or recessive. The relationship of the forms of colour is dominant to black and brown, and black to brown, therefore brown is recessive to them both. It is not correct to talk about the gene for red, black or brown, as genes can influence other characters, e.g. the gene determining brown pigmentation also influences the eye colour of brown individuals. Their eyes are the colour of a false pearl: they are never red or organic; and their plumage fades from exposure to sunlight.

Let us now examine the method by which the chromosomes and their genes are passed from one generation to the next. Fertilization takes place when a sperm of the male meets and fuses with an ovum of the female, forming the *zygote*, the first cell of the new individual. This first cell divides by a process termed *mitosis* which produces two daughter-cells identical with the one from which they were derived. This is accomplished by the chromosomes in the original cell splitting longitudinally into two halves called *chromatids*. After this has been accomplished each individual chromatid moves away from its other half. They now form two groups, each group containing two chromatids of each kind. A new nucleus membrane forms around each group. During this process the original cell membrane takes on an hour-glass shape, the neck narrows and finally breaks and the result is two new cells.

This is the method by which the body cells multiply to provide growth and the replacement of worn-out cells. At first, the body cells remain as they originally were, but gradually groups of cells begin to specialize in performing various functions. Those forming the nerves carry messages to and from the brain. Those of the muscles, by their power to expand and contract, bring about movement. The cells of the various organs and glands produce substances required by the body and/or remove waste matter; while those of the blood function as general carriers of nutrition and the removal of extraneous matter.

The cells in which we are interested have not altered from the original first cell (zygote) and are located in the *gonads*, i.e. the *testes* of the cock and the *ovary* of the hen. At the division of the body cells (mitosis) the new daughter-cells contain two chromosomes of every kind (diploid cells), but the cells of the gonads divide twice by a process called *meosis* producing *haploid* cells called *Gametes* containing only one chromosome of each kind. The gametes of the cock are called *spermatozoa* or *sperms* and those of the hen *ova*. These gametes, or reproductive cells, are capable of moving out of the body, those of the cock being very small chiefly because of the reduction of their cytoplasm, while those of the hen are comparatively large. Remember that they contain only one chromosome of each kind, but when a sperm and an ovum meet and fuse at fertilization, forming the zygote of the new individual, the numbers are brought up to two chromosomes of each kind; one inherited from the sire and the other from the dam.

For some time it was the concensus of opinion that hens only possessed a single *sex chromosome*. However, modern equipment and new techniques have disclosed the fact that this single-sex chromosome does have a partner. The sex which carries the pair of like sex chromosomes is termed the *homogamatic sex* and the sex carrying a pair of unlike sex chromosomes is known as the *heterogamatic sex*. One of the fundamental differences between pigeons and mammals, including man, lies in the fact that in pigeons it is the *male* that is the homogamatic sex and the hen that is the heterogamatic sex. In mammals it is the opposite: females are the homogamatic sex and males the heterogamatic sex. Recent research has revealed many other differences between the genetic constitution of pigeons and mammals, mostly too technical to be included in this chapter. One of these differences is that whereas the sex chromosomes in mammals are the largest, in pigeons they only rank fourth in size. To draw attention to these differences the sex chromosomes of pigeons are now denoted 'Z' and 'W'. Here it is important to remember that it is the 'Z' chromosome which carries the genes determining sex and all the other sex-linked characters. Those in which we are interested *dilute* (short down), and *colour* (pigmentation), all of which are visible to the eye; and there must be many more. The 'Z' chromosome's partner 'W' indicates femaleness but otherwise it is inactive in determining the characteristics carried on the 'Z' chromosome. This means that the genes carried on the single 'Z' of hens are said to be *hemizygous* for all sex-linked characteristics. Thus, whereas it requires two genes determining dilute to produce Duns and Silvers and Yellows in cocks, a single gene produces them in hens.

In order to differentiate between the sex chromosomes and all the other chromosome pairs, they are referred to as *autosomes* and the characteristics they control as *autosomal*. As both sexes possess the same number of pairs of autosomes the mode of inheritance of autosomal characteristics is simple and straightforward. They and their genes, at the formation of the gametes, segregate (Mendel's first law) and assert independently (Mendel's second law). This second law does not apply in cases of linkage.

The number of chromosome pairs in the body cells of different species differ: here is a random selection: fruit fly 4; pig 19;

man 19; horse 32; and examples of the Class Aves, sub-class Crinatae: parakeet 29; house sparrow 38; fowl (chicken) 78; canary 40; duck 40; pheasant 40; pigeon (Columba Livia) 40.

The situation with pigeons at the formation of the gametes is that every sperm of the homogamatic sex (cocks) contain 39 autosomes and 1 'Z' chromosome and the heterogamatic sex (hens) produce two kinds of ova: one kind containing 39 autosomes and 1 'Z' chromosome, the other kind contains 39 autosomes and 1 'W' chromosome with the result that the individual receiving it will be a hen.

The Inheritance of Sex

PARENTAL GENERATION P1

SIRE DAM

ZZ Chromosomes (sex) ZW

Gametes

Z Z sperms Ova Z W

Z Z Zygote 1st Filial Generation F1 Z W Zygote

Cock ZYGOTE (The first cell of the new individual) HEN

Figure 2

Note (*Fig*. 2) that the sire carries 2 'Z' chromosomes and the hen only 1 which she passes to her son and her 'W' to her daughter who receives her single 'Z' chromosome from her sire who in turn also passes a 'Z' chromosome to his son, with the result that we appear to be back to the same situation found in the parental generation. A tricky point arises concerning 2 'Z' chromosomes carried by the young cock of the first filial generation—one inherited from his sire and the other from his dam. When mated he will in his gametes (sperm) pass replicas of one or the other of these 'Z' chromosomes; which one is all a matter of the genetic chance. The one he passes in the first instance has no influence on the next one; there is no question of their being passed alternately (*see* note to *Fig*. 3).

When you ask a fancier the colour of any given pigeon he will answer Red Chequer Cock or Blue Chequer Hen and in doing so, whether he is aware of it or not, he is using three distinctive and independent genetic characteristics. Red and Blue tell us the form of pigmentation to which the individual belongs; chequering its plumage pattern; and cock and hen the sex. In order that there should be no misunderstanding I want to make clear that when I use the word *colour* I specifically mean 'pigmentation'.

The Inheritance of Colour

We pass now to the inheritance of colour which, as we found, it consists of three pigmentation groups, Red, Black and Brown forming an allelomorphic series which is carried on the 'Z' chromosome. Therefore, where a 'Z' chromsome goes, so also does a gene determining one of the three forms of colour and because of this colour is said to be *sex linked.*

To illustrate what happens we will mate a black cock which, having two 'Z' chromosomes, will carry two colour-determining genes; judging from his *phenotype* (an individual judged by its appearance) he must carry at least one gene determining black. But what does the other gene determine? It cannot determine red because then the bird would be a red: it could of course determine black, but we will assume it determines brown; in which case the black cock's *genotype* (the factors possessed by an individual, i.e. its genetic constitution), is black and brown.

When writing this, the dominant form of the character concerned is written in capitals above the line and the recessive form in small letters under the line, thus: $\frac{\text{BLACK.}}{\text{brown}}$ Since hens have only one 'Z' chromosome and therefore only one gene determining a form of colour they are written thus: $\frac{\text{RED.}}{\text{W}}$

Now the black cock is mated to a red hen.

Figure 3 illustrates how sex-linked characters are inherited, so we now pass on to the inheritance of autosomal characters. Here there is no difference between the sexes as they both possess thirty-nine pairs of autosomes.

PHENOTYPE				PHENOTYPE
Black Cock		(PARENTS)		Red Hen
Black	Brown	(GAMETES)	RED	W
RED / black	*RED* / brown	(GENOTYPES OF YOUNG)	*BLACK* / W/	*BROWN* / W/
Red Cock No. 1	Red Cock No. 2	(PHENOTYPES OF YOUNG)	Black Hen No. 3	Brown H3n No. 4

A colour denotes the presence of a 'Z' chromosome. W denotes its absence. Cocks carry two colours therefore two 'Z' chromosomes and hens only one.

Figure 3

Note how the red hen's single 'Z' chromosome, with its red determining gene, is passed to her sons and because of the dominance of red they are both reds. No. 1 carries black and No. 2 brown. When a sperm of the cock meets an ovum of the hen that does not carry a 'Z' chromosome the result is a hen, and her colour depends on which colour the cock's sperm carries—in this example either black or brown. The point to bear in mind is that red hens always produce red sons. How each individual pigeon will breed depends on its genotype: its phenotype is no certain guide.

Pattern

The designs of pattern are identical in each pigmentation group. In the red group the designs consist of red and ash-grey. The black group consists of black and dove-grey (or blue); and brown of brown and tawny.

There are five designs, namely, Dark Chequer, Chequer, Light Chequer, Barred and Barless. All alleles with darker designs are dominant to those lighter than themselves and vice versa; the lighter designs are recessive to all darker than themselves.

TABLE I

Designs of pattern	Forms of Colour		
	Red/ash Phenotypes	Black/blue Phenotypes	Brown/tawny Phenotypes
Dark Chequer	Dark Red Chequer	Dark Chequer	Dark Brown Chequer
Chequer	Red Chequer	Blue Chequer	Brown Chequer
Light Chequer	Light Red Chequer	Light Blue Chequer	Light Brown Chequer
Barred	Mealy	Blue	Brown Barred
Barless	Barless Mealy (True)	Barless Blue	Barless Brown

The fifteen phenotypes that the three forms of colour and the five forms of pattern can produce are those of normal plumage. Anything different from the above indicates the presence of an additional characteristic.

Distinguishing Features

RED/ash: ash-grey flights and tails. Heterozygous Red/ash cocks may have black splashes.

BLACK/blue: dark flights and a black terminal tail bar.

BROWN/tawny: brown flights and a dark brown terminal tail bar, with eyes a false pearl colour. Feathers fade from exposure to the sun.

Now let us mate a pair of pigeons, including in their genotype all the characters we have so far dealt with—colour, pattern and sex. Before doing so I would draw the reader's attention to the fact that when the pair of genes in a given allelomorph are similar they are said to be *homozygous* and the individual a *homozygote* for that form of a character, but if they are dissimilar they are said to be *heterozygous* and the individual a *heterozygote*.

The pair of pigeons chosen are both heterozygotes for pattern. A blue chequer cock heterozygous for barred and a light blue chequer hen also heterozygous for barred. Their respective phenotypes and genotypes are shown thus:

Phenotype	Genotype
Blue Chequer Cock	$\frac{\textit{BLACK/CHEQUER}}{\text{Black/barred}}$
Light Blue Chequer Hen	$\frac{\textit{BLACK/LIGHT CHEQUER}}{\text{W/barred}}$

Figure 4

Note that the dominant forms of the characteristics are in capitals above the line, and the recessive forms below the line, while the oblique stroke / separates the characteristics.

The cock can pass two kinds of gametes (sperm), black/chequer and black/barred, while the hen can pass in her gametes (ova) four kinds—black/light chequer; black/barred: W/light chequer; and W/barred. The first two carry a gene-determining colour and a 'Z' chromosome, while the last two do not carry a 'Z' chromosome and therefore no gene-determining colour. We now multiply the hen's gametes by those of the cock: $4 \times 2 = 8$, and draw 8 squares in 2 rows of 4 (*see* Table 2).

Table 3, in comparison, introduces the dominant red/ash form of colour. We will mate the Blue Chequer cock of Table 2 to a Mealy hen. The cock gives two kinds of gametes and the hen being homozygous for barred gives one carrying the 'Z' chromosome and with it a gene-determining colour. The other carries the inactive 'W' chromosome, denoting a female.

TABLE 2

Hen's gametes (ova)

Cock's gametes (sperms)	Black/Light Chequer	Black/Barred	W/Light Chequer	W/Barred
Black/ Chequer	BLACK/CHEQUER black/light chequer No. 1 Blue Chequer Cock	BLACK/CHEQUER black/barred No. 2 Blue Chequer Cock	BLACK/CHEQUER W/light chequer No. 3 Blue Chequer Hen	BLACK/CHEQUER W/barred No. 4 Blue Chequer Hen
Black/ Barred	BLACK/LIGHT CHEQUER black/barred No. 5 Light Blue Chequer Cock	BLACK/BARRED black/barred No. 6 Blue Cock	BLACK/LIGHT CHEQUER W/barred No. 7 Light Blue Chequer Hen	BLACK/BARRED W/barred No. 8 Blue Hen

This table demonstrates the expectancy this mating can produce. A study of the result discloses that there are eight different genotypes of the characters involved, yet there are only three phenotypes, blue chequers (4); light blue chequers (2); and blues (barred) with the sexes equally divided. Cock No. 2 is identical with his sire and hen No. 7 with her dam. The cocks are recognized by having two colours in their genotype, but then hens having only one 'Z' chromosome possess one gene-determining colour and for that reason are said to be *hemizygous*, Nos. 1, 2, 3, 4, 5, and 7 are heterozygous for pattern, with Nos. 6 and 8 homozygous for pattern (barred). Also homozygous for colour are all the cocks (Nos. 1, 2, 5 and 6). The blue cock No. 6 is the only complete homozygote for all the characteristics involved.

This table demonstrates the expectancy this mating can produce. A study of the result discloses that there are eight different genotypes of the characters involved, yet there are only three phenotypes, blue chequers (4); light blue chequers (2); and blues (barred) with the sexes equally divided. Cock No. 2 is identical with his sire and hen No. 7 with her dam. The cocks are recognized by having two colours in their genotype, but then hens having only one 'Z' chromosome possesses one gene-determining colour and for that reason are said to be *hemizy-*

gous. Nos. 1, 2, 3, 4, 5, and 7 are heterozygous for pattern, with Nos. 6 and 8 homozygous for pattern (barred). Also homozygous for colour are all the cocks (Nos. 1, 2, 5 and 6) The blue cock No. 6 is the only complete homozygote for all the characters involved.

TABLE 3

Blue Chequer Cock × Mealy Hen

Cock's gametes (Sperms)	Hen's gametes (Ova)	
	Red/Barred	W/Barred
Black/Chequer	RED/CHEQUER black/barred No. 9 Red Chequer Cock	BLACK/CHEQUER W/barred No. 10 Blue Chequer Hen
Black/Barred	RED/BARRED black/barred No. 11 Mealy Cock	BLACK/BARRED W/barred No. 12 Blue Hen

Table 3 confirms that red/ash hens always produce red/ash sons, and that hens inherit their pigmentation group from their sire. Another point of interest is that in this family of six pigeons the gene distribution is as follows: 9 determine the barred pattern; 6 Black pigmentation; 3 the Red; 3 the Chequer pattern, giving a gene-frequency ration of 3—2—1—1. Yet owing to the dominance of Red/Ash and Chequer there are six different genotypes and phenotypes.

In Table 4 we will reverse the sexes regarding pigmentation by mating Red Chequer Cock No. 9 and Blue Hen No. 12 from Table 3.

Red and Blue Chequers, Mealies and Blues are in equal numbers of both sexes. Again the gene-frequency is in the ratio of 3—2—1—1, that is 3 barred; 2 black; 1 red and 1 chequer.

With random matings the gene frequency within a family of pigeons the gene ration remains constant. However, with careful selection you can alter the frequency by eliminating the characters not desired. The elimination of the dominant characters is relatively easy. All you need do is to stop breeding

TABLE 4

Red Chequer Cock × Blue Hen

Hen's gametes (ova)	Cock's gametes (sperms) Red/Chequer	Black/Chequer	Red/Barred	Black/Barred
Black/ Barred	RED/CHEQUER black/barred No. 13 Red Chequer Cock	BLACK/CHEQUER black/barred No. 14 Blue Chequer Cock	RED/BARRED black/barred No. 15 Mealy Cock	BLACK/BARRED black/barred No. 16 Blue Cock
W/Barred	RED/CHEQUER W/barred No. 17 Red Chequer Hen	BLACK/CHEQUER W/barred No. 18 Blue Chequer Hen	RED/BARRED W/barred No. 19 Mealy Hen	BLACK/BARRED W/barred No. 20 Blue Hen

from the birds possessing them. The recessive forms are difficult to eradicate and birds known to possess them must be eliminated and also their parents and sibs. In the selection of mating birds it is obviously important to know their respective genotypes. Their appearance or phenotype is no sure guide as to how they will breed, it being only half the picture. The other half is hidden by dominance. If we know how the parents are bred we can make only a rough guess at what is probably hidden. The only sure way of knowing is to breed from the individual concerned, when mated to a bird of whose genotype we are sure, e.g. Blues.

Every pigeon inherits and carries in its genotype the genes determining one form or another of these three visible characters, colour, pattern and sex. There are five forms of pattern, three forms of pigmentation and two sexes. The first two can combine in fifteen different ways to produce fifteen different phenotypes as listed in Table 1.

Some pigeons while inheriting the characters in Table 1 inherit additional characters which in different ways alter the genotype and phenotype of these normal plumaged pigeons listed in Table 1. Irrespective of what the finished article may look like in its phenotype, always remember that it carries in its genotype (i.e. genetic constitution) the genes determining normal plumage.

Additional Characters

These additional characters can be roughly divided into two types: *Epistasis*, where a character which is not an allele masks the effect of another character or characters. The most common is *self colour* or *spread*. It is dominant to its allele normal plumage; its effect is *epistatic* masking all forms of pattern. It turns all birds of the red/ash group to Ash Reds (commonly called Barless Mealies), black/blue group to Self Blacks and brown/tawny to Self Browns. Some confusion exists from the assumption that blue is a basic colour when in fact it is only a derivature of black pigmentation, as ash is to red and tawny is to brown.

Recessive Red, like self colour, is an autosomal character, but as its name implies it is recessive to normal plumage. In its homozygous state it masks all pattern and colour, turning black/blue and brown/tawny birds into Brick Reds. With the dominant red/ash birds when homozygous it turns them into Brick Reds. They are distinguished by the fact that their flights and tails are red to the tips. It is epistatic in its effect masking both colour and pattern. I have observed two peculiarities: first that some recessive reds have odd feathers of another colour in their tails, and that in its heterozygous state (only one gene determining it) with red/ash birds it turns them into a muddy clay shade of red including flights and tails.

White is a bit of a mystery because there is both a dominant and a recessive type, both epistatic masking all colour and pattern. Despite their white plumage both types carry the full complement of genes determining the forms of colour and pattern of normal plumage. Whites, when mated to normal plumaged birds, produce a surprising variety of colour and patterns in their progeny. Most of the whites have bull eyes but some have the normal run of coloured eyes. These variations observed when mating whites to normal plumaged pigeons are the result of the genes determining normal plumage is giving expression when free from the epistatic effect of white.

Pieds range from birds with a few white feathers about the head and neck and/or a few white flights to the gayest of gay pieds. The progeny of two birds with white feathers usually have more white feathers than their parents. Some gay pieds are almost completely white, except for their wing covert feathers which display (turbit-like) one of the five forms of pattern.

The second type of additional characters are those I describe as 'decorative', because their effect is superimposed on the normal plumage but not masking it as with epistasis.

Earlier on I stated that 'Every pigeon inherits and carries in its genotype one form or the other of these three visible characters: colour, pattern and sex.' The three epistatic characters *self colour*, *recessive red* and *white* illustrate what this means. Mate a self Black or Ash Red (so called Barless Mealy) to a normal plumaged pigeon as listed in Table 1 and on the average the progeny of such a mating will produce 50 per cent Self Blacks or Ash Reds and 50 per cent normal plumaged pigeons. Mate a recessive Red to a normal plumaged pigeon and the recessive red character will disappear despite the fact that 50 per cent of the progeny will carry a single gene-determining recessive red now masked by its dominant allele normal plumage. When two birds carrying the masked gene determining recessive red the progeny will be recessive reds, much the same applies to whites.

Grizzling is an autosomal character, and a simple dominant to normal plumage. Its white flecking on head and neck is well known. Only a single gene determining grizzle is required for it to be expressed and in this heterozygous state there is no difficulty in recognizing the colour and pattern of the bird concerned. When homozygous, the pattern may be uncertain, but the colour is always apparent. There must always be a grizzle in a mating to produce grizzles, repeated grizzle to grizzle matings tend to produce progeny that are almost completely white. The plumage of grizzles become darker as the bird gets older.

Dapple (Hollander's Sooty) is an autosomal decorative character which when present superimposes its effect on all normal plumaged pigeons, easily recognized in Barless Blues, Blues and Strawberry Mealies respectively. It is thought by many to be recessive because of its sudden and unexpected appearance in these Pencil Blues and Strawberry Mealies. But in fact it resembles chequering in many respects and appears to consist of a series of multiple alleles ranging from a few hazy chequers on their lesser ceverts to a great number of these hazy chequers as found on well-marked Pencil Blues and Strawberry Mealies. Dapple is present in many Chequers if you know where to look and what to look for. The difference in the markings lies in the fact that the chequering markings spread from the outer edge of the feathers inwards and are of intense shade of black whereas those of dapple are at the tip of the affected feathers spreading from the quill downwards and of a lighter shade. Both characters are expressed on the coverts and lower edges of the wing bar. When both these characters are expressed in the same pigeon they have the effect of darkening the plumage so that a Blue Chequer may look like a Dark Chequer, this applies to all the colour groups. Sometimes dapple is not expressed in the nest feathers only to appear at the first moult.

Opal is an autosomal character, recessive to normal plumage and known to racing pigeon fanciers as *mosaic*. Its characteristic effect is to bleach out parts of the dark areas of the feathers, leaving dark edges in black/blue or red/ash best seen in the darker forms of pattern. *Dilute* is a sex-linked character recessive to normal plumage. It changes the colour not the pattern. Blue chequers and light blue chequers change to *silver chequers* and blues to *silvers*. Hens are *hemizygous* (having only one 'Z' chromosome, the genes carried on it are not paired) and the presence of a single gene has the same effect as two with cocks. With recessive reds and red/ash birds it turns them into *yellows* and the browns to *khaki*. The gene which determines dilute has another effect in that the newly hatched youngsters have short down and look naked by comparison with others. When a short downed dilute youngster turns up from normal plumaged parents it is always a hen.

Smoky is an autosomal character recessive to normal plumage. It darkens the plumage and very noticeable in blues and light blue chequers but effects all other patterns to self blacks.

Birds with smoky can be spotted by the fact that the base of their beaks are of a light colour.

Frill and *Feather Leg* recessives which occasionally turn up as a reminder of characteristics possessed by a far back ancestor.

At the first time of reading, all this may appear complex and difficult, but with a little patience and progressing step by step, your interest will grow and the difficulties disappear. I know of no better way of familiarizing yourself with the operations of the laws of inheritance, the meaning and use of the genetic terms than to take pencil and paper and work out the expectancy of the matings in your own loft. Think of the interest engendered by checking them against the actual results obtained. The first step is to write down the genotype of the partners to the mating. Next find the number of kinds of gametes each partner can pass, multiplying the one with the other to find the number of squares required. When using the three characters, colour, pattern and sex, the number of squares required can be 4, 8 and 16 for each additional character these numbers double up. Follow the procedure as in Tables 2, 3, 4 and 5. Remember that it is the dominant form of the character that is shown in capitals above the line and its recessive allele below the line in small letters. Those above the line are the individual's phenotype and the combination of characters or forms of characters above and those below the line are its genotype. The individuals with two colours are the cocks and those with only one colour above the lines are the hens.

When an individual in a mating possesses one of the additional characters this has to be added and the number of gametes it can pass doubles up. In order that a comparison can be made we will assume that the cock is a Blue Chequer Grizzle and the hen a Mealy. This is the same mating as in Table 3 only we have added grizzle to the cocks phenotype and genotype. This means that he can pass four kinds of gametes (in Table 3 he only passed two) and the hen's gametes remain as four (the small 'g' in Table 5 denotes the absence of grizzling). Their respective genotypes are:

TABLE 5

Blue chequer grizzle — Mealy hen

GRIZZLE/BLACK/CHEQUER × RED/BARRED
g/black/barred — W/barred

Cock's gametes	Blue Chequer Grizzle × Mealy Hen Hen's gametes Red/Barred	W/Barred
Grizzle/ Black/ Chequer	GRIZZLE/RED/CHEQUER g/black/barred No. 21 Red Chequer Grizzle Cock	GRIZZLE/BLACK/CHEQUER g/W/barred No. 22 Blue Chequer Grizzle Hen
Grizzle/ Black/ Barred	GRIZZLE/RED/BARRED g/black/barred No. 23 Mealy Grizzle Cock	GRIZZLE/BLACK/BARRED g/W/barred No. 24 Blue Grizzle Hen
g/ Black/ Chequer	g/RED/CHEQUER black/barred No. 25 Red Chequer Cock	g/BLACK/CHEQUER W/barred No. 26 Blue Chequer Hen
g/ Black/ Barred	g/RED/BARRED black/barred No. 27 Mealy Cock	g/BLACK/BARRED W/barred No. 28 Blue Hen

The expectancy in this table gives us 50/50 grizzles and normal plumaged pigeons and you will find that the bottom four squares are the same as in Table 3. If we substituted self colour for grizzle we would get 50/50 self coloured and normal plumaged pigeons. Nos. 21 and 23 would be Ash Red cocks (Barless Mealies, so called). Nos. 22 and 24 would be self black hens. The others would be normal plumaged.

A close study of Tables 2 and 5 inclusive and the working out of a few of your own matings and the results obtained will disclose the shuffling of the genes that takes place at the formation of the gametes and that the progeny rarely resemble the parents or each other in every respect. Any likeness between parents and *sibs* (progeny of the same parents) is understandable as they have been derived from the same hereditary material. The individual chromosomes need not have changed in any way: the changes we observe are brought about by the shuffling of the genes which are allelomorphic to each other (alleles).

Mutation

Mutation is a change in any unit of heredity. The majority of these are of individual genes, in which a gene, while still influencing the same character as before, does so in a different way. If the gene which has hitherto determined the barred form of pattern changes and now produces a pigeon still with bars, but in addition has a few dark chequers on the covert feathers, it will be a Light Blue Chequer. It is easy to see from this how and why multiple alleles arise. A mutant gene is just as stable in its new form as it was in its old. Mutations are extremely rare.

Linkage

You will recall that the characters colour and dilute are carried on the 'Z' chromosomes together with sex are said to be sex-linked. There must be many linkages carried on the autosomes. There is a tendency for genes to segregate together instead of assorting independently because they are carried on the same chromosome or autosome. Dr. W. F. Hollander of Iowa State University found that a linkage exists between 'self' or 'spread', 'pattern' and 'opal' as they are carried on the same autosome. 'Pattern' and 'opal' are situated close to each other while 'self' is more or less situated at the other end of the autosome. Characters which are located close together are said to have a strong linkage while those situated some distance apart are said to have a weak linkage. The reason for this is that, by a process termed *crossing over*, pairs of homologous chromosomes exchange parts during their pairing at meiosis. To illustrate this I will represent a pair of homologous chromosomes by a series of letters: those in capitals have come from the sire and those in small letters from the dam.

The pairs of letters represent an allelomorph controlling character.

SIRE: A.B.C.D.E.F.G.H.I.J.K.L.M.N.
DAM: a.b.c.d.e.f.g.h.i.j.k.l.m.n.
The matched chromosomes before crossing over.

SIRE: A.B.C.D.E.F.G.H.I.J.k.l.m.n.
DAM: a.b.c.d.e.f.g.h.i.j.K.L.M.N.
The same chromosomes after crossing over.

Figure 5

Note: That while there has been an exchange of the letters K.L.M.N. and k.l.m.n. between the chromosomes, all that has happened is the breaking of the original linkage and the creation of a new one. The genes affected have changed chromosomes but they still control the same characters as before. The chromosome which originally carried the gene determining the dominant form of a character now carries its recessive and vice versa. Where the genes are similar there has been no apparent change.

A little thought on the situation illustrated in Fig. 5 will make it clear that genes situated close together are less likely to be separated by a cross-over between them but that the further any two genes are apart the greater are the chances of them being separated by a cross-over. Such changes make a difference to the gametes the partners to a mating can pass to their progeny.

A personal experience of this was in the mating of a Self Black Cock to a Dark Chequer Hen: for the first two seasons they produced Dark Chequers and Self Blacks, but in their third season they produced Self Blacks, Dark Chequers and Blues. Obviously a cross-over had taken place between self and pattern. For the first two years the cock's gametes carried self/barred and s/dark chequer and the hens carried dark chequer and barred. The cross-over changed the cock's gametes to self/dark chequer and s/barred but the hen's gametes remained dark/chequer and barred. In the first two years the epistasis of self and the dominance of dark chequer masked the barred pattern although both parents carried it in their genotype. After the cross-over self's epistasis changed from masking barred to masking dark chequer with the result that barred pattern was able to be expressed. Table 6 illustrates the differences brought about by the cross-over.

TABLE 6

COCK'S GAMETES

	Before crossing over		After crossing over	
HEN'S gametes	SELF/barred	Dark Chequer	SELF/dark Chequer	Barred
Dark chequer	SELF/barred s/dark chequer Self Black	DARK CHEQUER dark chequer Dark Chequer	Self/dark chequer s/dark chequer Self Black	DARK CHEQUER barred Dark Chequer
Barred	SELF/barred s/barred Self Black	DARK CHEQUER barred Dark Chequer	SELF/dark chequer s/barred Self Black	BARRED barred Blue

The small 's' denotes normal plumage, the recessive allele of self. The form of pattern on the same chromosome as Self is shown in small letters indicating that it is masked.

Improving Your Stock

The changes brought by crossing over plus the shuffling of the genes at the formation of the gametes of both partners in a mating account for the variations with the visible characters we have been studying. Watching the operation of the laws of inheritance among our own pigeons, we become better equipped to tackle the problem of improving our stock.

In the average loft without any definite plan of selection the gene frequency in a family of pigeons remains more or less constant from one generation to another. Surely it is the desire of every fancier to improve the quality of his family. This can be done by careful selection in the choice of mated pairs, each partner having the desirable characters and, as far as we can judge, no unwanted characters. Gradually the proportion of genes determining the desirable increases and, at the same time reduces those genes determining the unwanted characters, thus changing the gene frequency in the family.

The elimination of dominant characters is comparatively simple—all you need do is not to breed from them. Getting rid of the undesirable recessives is not so easy. Turn to Table 2. In this family of ten pigeons we have four Blue Chequers and two Light Blue Chequers and two Blues (barred). Assuming the barred pattern is not wanted, getting rid of the two Blues is not enough, because all the other eight carry a gene-determining barred. To eliminate barred we would require to stop breeding from all ten of these pigeons and their grandparents and any others bred from them. The reduction of the genes determining unwanted characters results in an increase of their desirable allelomorphs.

The majority of fanciers are guided by ancestry and breed mostly within a family related to a famous ancestor, but the value of this method decreases rapidly as the nearness to the famous ancestor and the birds being bred from decrease. Inbreeding is the mating of close relatives and it increases the number of characters in which the genes determining them are homozygous. With every generation of continued inbreeding the number of individuals and characters which become homozygous increases; this situation is termed *homozygosity*. Desirable as homozygosity is for the desired characters, the same process which produces it also brings the unwanted characters into the homozygous state.

The mating of close relatives, simply because they are closely related, is not enough. Great care must be taken that they possess the desirable characters and equally important that they are free of any unwanted characters. Therefore, every daughter is not genotypically a suitable mate for her sire, or son for his dam. If we are to obtain the desired result, every mated pair must carry the genes determining the characters we want and under no circumstances an undesirable one.

Every pigeon, before it can produce, must have a mate and each contributes part of itself to their progeny. We have ample evidence of the variations which occur in most matings: homozygosity tends to reduce this variation and as it spreads in a family the members become increasingly uniform of type.

Fanciers, in their enthusiasm, use many superlatives when describing the virtues of their favourites but most of these would be difficult to substantiate. I have seen and handled many champions, and by champions I mean pigeons which have flown and won over and over again from 500/600 miles.

Regarding their physical features, I found that in the hand they gave me the impression of being alive and buoyant, possessed of fine tight-fitting feathers and, with the odd exception, they had no exaggerated features and each part was in keeping with their overall size, large or small, in other words balanced. Not one of them resembled the over-feathered, over-bodied, heavy-as-lead specimens I have met with when judging. Weight is a handicap in a racer or show-bred pigeon.

For a number of years now I have kept note of how these good pigeons were bred and the breeding and performance of their near relatives; how they were handled in management by their owners; the state of their moult and nesting conditions at the time of basketing, when successful; also their mistakes and possible reason for them. All this was observed in my search for what champions had in common—seen and unseen. My conclusions are: obtain sound vigorous pigeons possessing in full measure all the natural habits of the species; pigeons that are never sick or sorry; that mate, go to nest, rear their youngsters without turning a feather and go to nest again just as keenly as they did the previous nests. If they don't behave thus there is something wrong with them, or your management of them.

The right kind need no coddling but take all you ask of them in their stride. What to breed for is *stamina*, *vitality* and *fertility* and what better proof that a pigeon has these than *longevity*? It is positive proof of a sound constitution, resistance to disease, vigour, vitality and stamina. These are the inherited characters which combine to produce a champion; they cannot be seen but can be recognized by probing an individual's family history and its own performances as a racer and producer. This kind of pigeon is most likely to be found in a loft winning now (not a loft with an isolated champion) with a cosely-knit family, the members of which have for generations raced and won from the extreme distances over and over again and lived to be old pigeons and produced their kind right up to the end.

In selecting pigeons to mate, avoid mating extremes, thinking the one will balance the other out; it is much safer to mate two of a similar type, physically, always providing they each possess to the best of your knowledge all the essentials listed above. Remember that if one of the partners has a weakness, it may not show up in the first generation, hidden by a dominant allele of the other partner, but you are only perpetuating it in the family and it will show up sooner or later.

In mating a pair of pigeons, each with desirable qualities, the nearness of relationship does not really matter. However, if these aims can be produced by inbreeding, so much the better, since it reduces the chances of an undesirable character turning up. Similarly, if one is going to appear it is likely that it will become apparent sooner and can be dealt with at once, thus saving valuable time. Elimination is the key to this purifying process.

There is always danger in introducing a new pigeon to your loft: it may introduce an unexpected, unwanted character. The progeny from any introductions must be severely tested before breeding freely from them. A high percentage of the champions of which I have detailed records, and there must be many hundreds more, have been the result of *heterosity*, perhaps better known as *hybrid vigour*. It results from the mating of two unrelated but inbred pigeons (first cross), provided that the partners to the mating are from two families known to possess the essential qualities already mentioned. It is likely that they both will be homozygous for some of the essentials and will breed true for them. In addition to this each may well introduce something that the other hasn't got, and they will dovetail, or as we say 'nick'. Then the progeny will be more vigorous than either parent and likely to produce the outstanding champion.

One must always bear in mind that is the respective genotypes of the mated pair which have nicked, being complementary to each other and not necessarily their respective families; even brothers and sisters of the partners when mated together may not nick, but it is worth trying. It is not unusual to find that champion racers, the result of a first cross, don't turn out to be successful producers.

Examples of famous champions, the outcome of heterosity with family histories in which stamina, vigour, virility and longevity were prominent features, will be found ranging from S. P. Griffiths' '459' bred in 1908 to the present day;'459' was bred from inbred but from unrelated parents. In her day she was considered 'the best pigeon the fancy had ever known'. A winner at every stage from Bath to Marrennes, 530 miles, she startled the pigeon world by winning the Great Northern Marrennes race for a second year in succession. In addition to winning a record sum in prizes and pools for those days she also won a motor car.

A more recent example is H. J. King's *Twilight*, bred in 1954. In N.F.C. races she won 4th Sect/8th Open; 3rd Sect/4th Open; from Pau, 6th Sect/59th Open Bordeaux and then at age five 1st Sect and 1st Open King's Cup and many other trophies from Pau. She was bred from inbred but unrelated parents. Now watch this: her sire 2423 was aged five when he produced her. 2423's sire *Champion 91* was aged fifteen when mated to his own granddaughter *Feardale Queen* aged five when they produced 2423. *Twilight*'s dam 1477 was aged six when she produced her; 1477 was bred from *Newmarket Queen*, aged ten, when mated to her own yearling son. Going back twenty years in the pedigree of *Twilight* there is ample evidence of stamina, vigour and longevity.

These are the inherited characters which every champion has in good measure, but do not overlook that every pigeon is as much the result of its environment as its heredity. A pigeon's environment consists of the air it breathes, the kind of food and water it consumes, its exercise and training; in fact everything which effects its every day life. In the best environment inherited characters are given every opportunity to develop fully, whereas in a poor environment they are repressed and stilted. There are many potential champions that never get the chance to prove themselves. The champion pigeons we know were kept racing year after year and not kept to look at, after a single good performance.

I have written the facts as I know them and hope the reader finds them not only interesting, but informative and helpful. I am only too well aware that it is next to impossible to master the intricacies of this subject without thoughtful study, particularly with your own pigeons. The making of an extra effort to do so will be well worthwhile, creating additional interest in the results of your own matings, stimulating your powers of perception with the result that visiting shows or other lofts you notice details that otherwise would have passed you by.

TABLE 7

COLOUR MATING

Genotypes Cocks		Hens	Colour and sex of young
RED			
Red	with	Any colour of hen	Reds of both sexes
RED			
Yellow	„	red, black or brown	Red cocks and red and yellow hen
„	„	yellow, dun or khaki	Reds and yellow of both sexes
RED			
Black	„	red or yellow	Red cocks, red and black hens
„	„	black or dun	Reds and blacks of both sexes
„	„	brown or khaki	Reds and blacks of both sexes
RED			
Dun	„	red	Red cocks, and red or dun hens
„	„	black or brown	Red and black cocks, red and dun hens
„	„	yellow	Red and yellow cocks, red and dun hens
„	„	dun or khaki	Reds and duns of both sexes
RED			
Brown	„	red or yellow	Red cocks, red and brown hens
„	„	black or dun	Red and black cocks, red and brown hens
„	„	brown or khaki	Reds and browns of both sexes
RED			
Khaki	„	red	Red cocks, red and khaki hens
„	„	black	Red and black cocks, red and khaki hens
„	„	brown	Red and brown cocks, red and khaki hens
„	„	yellow	Red and brown cocks, red and khaki hens
„	„	yellow	Red and yellow cocks, red and khaki hens
„	„	dun	Red and dun cocks, red and khaki hens
„	„	khaki	Reds and khaki of both sexes
BLACK			
Black	„	Red or yellow	Red cocks, black hens
„	„	black or dun	Blacks of both sexes
„	„	brown or khaki	Blacks of both sexes
BLACK			
Brown	„	red or yellow	Red cocks, black and brown hens
„	„	black or dun	Black cocks, black and brown hens
„	„	brown or khaki	Black and browns of both sexes
BLACK			
Dun	„	red	Red cocks, black and dun hens
„	„	black or brown	Black cocks, black and dun hens
„	„	yellow	Red and yellow cocks, black and dun
„	„	dun or khaki	Blacks and duns of both sexes
BLACK			
Khaki	„	red or yellow	Red and yellow cocks, black and khaki hens
„	„	black or dun	Black and dun cocks, black and khaki hens
„	„	brown	Black and brown cocks, black and khaki hens
„	„	khaki	Black and khaki of both sexes
BROWN			
Brown	„	red or yellow	Red cocks, brown hens
„	„	black or dun	Black cocks, brown hens
„	„	brown or khaki	Browns of both sexes
BROWN			
Khaki	„	red or yellow	Red and yellow cocks, brown and khaki hens
„	„	black or dun	Black and dun cocks, brown and khaki hens
„	„	brown or khaki	Brown and khaki of both sexes
YELLOW			
Yellow	„	red, black or brown	Red cocks and yellow hens
„	„	yellow, dun or khaki	Yellows of both sexes
YELLOW			
Dun	„	red, black or brown	Red or black cocks, yellow and dun hens
„	„	yellow	Yellow cocks, yellow and dun hens
„	„	dun or khaki	Yellow and duns of both sexes
YELLOW			
Khaki	„	red, black or brown	Red, black or brown cocks, yellow and khaki hens
„	„	yellow	Yellow cocks, yellow and khaki hens
„	„	dun	Red and black cocks, yellow and khaki hens
„	„	khaki	Yellow and khaki both sexes

Red in the above table is the Dominant Red (Ash). Duns include silvers and yellow includes the so-called creams.

10 The Proof of the Pudding

Royal Road

Pigeon racing is a strenuous sport; the path of success, like true love, never runs smooth and our few cherished successes are made all the sweeter by our many disappointments. It is not my intention to tell anyone how he should race and manage his pigeons. What I have written in the foregoing chapters is a sincere attempt to convey to the reader my considered conclusions and accumulating knowledge concerning racing pigeons based on my experiences from 1909, when I first put a ring on a youngster, until the present day. I am thoroughly convinced that there is no royal road to success and that the methods adopted with apparent success by one fancier may not work if attempted by another. We all possess different temperaments and since our hours of work and leisure differ we must all work out the plan of management best suited to our individual circumstances.

Competition Makes the Man and Birds

Our loft location in relation to other competitors plays an important part in the kind of fancier we become. By this I mean that some of the best long-distance racers have been hopelessly placed as far as ordinary club and Federation racing is concerned with the result that they have been forced to concentrate on the long-distance events. There is no doubt in my mind that the kind of competition in which a fancier finds himself does influence his outlook and the keener the competition the better the man and his birds. Hundreds of fanciers passed through the Pigeon Service Holding Unit while I commanded it during World War II; men from all parts of the British Isles found their way there and underwent an entrance test before being accepted as suitable. We soon found, however, that with few exceptions the best men came from certain areas of the country where the competition in racing was the keenest. Mentioning tests bring to mind some amusing incidents in connection with them. In the very early days of the pigeon service for both the Army and the RAF the numbers required were comparatively small. Fanciers were requested to make application and in due course were called up for interview and the numbers required selected. Later on, men for the Army Pigeon Service were drawn from men already in the Army and the procedure was for a soldier to make application to his Commanding Officer for transfer to the Army Pigeon Service. If this was granted he was posted to the PSHU where he was interviewed and if found suitable transferred to the Royal Corps of Signals. It soon became apparent to me that Commanding Officers very naturally were reluctant to let a good soldier leave his unit but only too willing to let others go.

Rough but Effective Test

One day I received information that a number of men of a certain unit had made application for transfer. As it happened I was to be in the vicinity of this unit a few days later so I arranged to interview these applicants. Thinking that there would be only a few, I allowed myself an hour to do the job before keeping another appointment. Imagine my surprise and dismay on arrival to find fifty or sixty men all lined up ready to be interviewed. The first man in came from around Airdrie. I forget his name, but it was soon evident that he knew the business. The next two floundered badly and it was clear to me that I was going to be there for hours. Then I had an idea. In my pocket I had a pigeon diary containing the photographs of four prominent winners of the previous year. They were a Red Cock and a Red Chequer, Blue Chequer and Blue Hens. Handing the four photographs to each man as he came in I asked him to tell me their colour and sex. What I remember most about this incident was the laughter of the Airdrie fancier at the description of these pigeons given by the majority of the men interviewed. The men who described them as greys and browns got short shrift and those who managed the colour stumbled over the markings and sex, with the result that in no time I had reduced the likely applicants to half a dozen for further questioning.

Another time, on a similar mission, I was accompanied by a senior officer who knew very little about pigeons. I showed him how quickly the photograph test separated the wheat from the chaff and remember his delight as he sorted them out leaving me to question further those he passed by the photograph test. True, it was a rough and ready test but in the circumstances very effective. I have interviewed men who had kept pigeons for years and were actually members of a club but who could not tell how long eggs took to incubate or the number of feathers a pigeon had in its tail.

Soldier First

The laugh was not always on the applicant for transfer. I recall Bill Hammond coming into my officer one Friday night about 5 p.m. telling me that a new man had arrived and would I see him. When a very smart soldier in the uniform of a famous regiment came through the door and saluted, I was impressed because one of my difficulties was to drive home to the men that they were soldiers first and pigeon fanciers second. After asking this soldier a few questions I was about to dismiss him when he asked if he could have a leave pass for the week-end. At that particular time the fear of invasion was at its height and only a small percentage of men were allowed on leave at any one time. I had just finished signing the leave passes and had noted I was over my quota and I therefore turned his request down. On Saturday night, Sergeant Stockdale reported that this man was missing but he turned up on Sunday night. On Monday morning he was up before me on the charge of being absent without leave and, being doubtful of his pigeon knowledge, without more ado I sent him back to his unit as unsuitable. A few days later when at the War Office, trouble came my way. It appears that this man shared a billet with some of the Army Pigeon men attached to our branch at the

War Office and had become friendly with them. How or why I do not know, but the facts were that this man had not asked for a transfer in the usual way but our branch at the War Office had requested his transfer. The reader can well imagine the hubbub my sending him back to his unit created. Anyway, I stuck to my plea that I acted in the best interests of the old standby of 'good order and military discipline'. Fortunately, I did not put in my report that he was not a good pigeon man as I found out later that someone in authority had tested and passed him as being competent. After it was all over Bill Hammond, Arthur Stockdale and I had a good laugh over it, but it made us wary and we took pains thereafter to discover from what source each man's application for transfer originated. Since the end of the war I have found that many men we could have done with were refused permission to transfer by their Commanding Officers.

Knowledge is Power

The harder the going and the keener the competition the better the men and the birds produced. In the long run we only get out of the sport just as much thought, time and hard work as we put into it. One careless thought or action can spoil the good work of a whole season's racing. Factual knowledge is power, and nothing, no matter how small or apparently trivial, concerning our pigeons should be allowed to pass without being investigated, though care must be taken to keep it in its proper perspective. Each little incident has a message for us if we are prepared to consider it from every angle in an attempt to discover its cause. We must check and counter-check before coming to any definite conclusion; in other words think again, and again. The good fancier, like every good craftsman, is not only particular about the material he works with but takes pride in how he handles it. Keeping a team of pigeons going from the start to the finish of a racing season calls for careful planning and skillful management. In no aspect of our sport is the skill of a fancier seen to greater advantage than in his preparation for and handling of the unforeseen emergencies. The first pigeon which fails to return from a training toss or race sets the ball rolling by presenting us with the problem of what to do about its mate. Should we try to keep it sitting for a day or two in the hope that the lost bird will return, or leave it to its resources?

The Lost Pigeon

My experience has been that if you do not get a pigeon by the third day it is not likely to turn up for a week to ten days. How to keep a mateless pigeon sitting depends on several circumstances. There is the pigeon itself, how long it has been sitting, and what steps you have taken in order to help it, right from the start. Like the ritual I employ when feeding, I have one which I follow strictly to the letter when my birds are sent racing. The nest boxes of every pair I sent racing are immediately closed when I basket them. This prevents their nest boxes being claimed by other birds left in the loft. When one of a pair sitting on eggs is left behind in the loft, they are taken away from the nest box during the period of the day when normally their mates would be sitting. That is cocks are removed between 6 p.m. and 9 a.m. and hens between 9 a.m. and 6 p.m. I put them in another compartment or just basket them. When I allow these birds back to their nest boxes I take care that their eggs are at least warm, for you must remember that a sitting pigeon has its skin against the incubating eggs. Once racing has started I take care to mark each egg as it is laid, with the date and the pair number. This prevents confusion arising when one commences shifting eggs around when basketing for a race or on the return of the race birds.

Small Youngsters

It is rare to find a pigeon left to attend to a small youngster giving any trouble, but I still like to give them a spell off from their parental duty by putting the youngsters under another bird for an hour or two. I would consider it nothing short of a disaster to find that I had not at least a pair of small youngsters of less than a week old in the loft at any given time during the racing season. When mating up in the spring I deliberately stagger my matings so that I do not find my birds all in the same nesting condition at one and the same time. This means that as each marking night comes round I have birds in almost every nesting condition: driving cocks, pigeons of both sexes sitting less than twelve days, others with eggs chipping or youngsters newly hatched, which come in useful to slip for an hour or so before basketing under the birds sitting around sixteen or seventeen days on eggs. Small youngsters come to my rescue in an emergency in very many ways. For instance, to keep happy pigeons which you are trying to keep sitting while their mate is away racing. Similarly, the returned racer can be induced to sit by being given a small youngster. When it has settled, the youngster can be taken away and the bird concerned will usually continue to sit on the eggs alone. Many a time I have had one partner to a mating sitting on eggs and the other on a small youngster. This is done by keeping the partners apart, except at the change over when you switch the eggs for a youngster and vice versa. I have on record the case of a late bred hen carrying four nestflights which raced six successive weeks in the following nesting conditions. First week, sitting two days; second week, sitting nine days; third week, sitting sixteen days; fourth week, sitting twenty-three days (a youngster was slipped under her just prior to basketing). At this stage her mate, an old cock, became restless and in order to keep him sitting he was given a youngster. When she returned from the race she took the eggs, but the cock would not sit without a youngster so I kept switching the eggs and youngsters as each took their turn of sitting, until the day before basketing when the youngster was left in the nest for the hen. She has then been sitting thirty days when sent to the fourth race and had been feeding the youngster one week when basketed for the sixth race, 250 miles. She was timed in the fifth and sixth race being my third bird to the loft on each occasion. When she was basketed for the sixth race, I took the youngster away and got her down on eggs again. She was sent sitting thirteen days to a 300-mile race and the following week sent hatching she successfully flew 360 miles on a hard day. While checking the above details I noted that another late bred, a cock this time, also carrying four nest-flights was my second bird in the fifth race winning 5th club sent driving. He was second bird from 365 miles in a north-west wind, velocity 910, winning 6th Club sent sitting on the third nest two days. Both these latebreds had a night out in the first race, and because of the circumstances I might have been excused racing them any further that season. The hen, 876, was specially bred for me by A. Pickering of Hinckley and the cock, *600 Tenacity*, was the youngster I had from *Tutimes* and *Fidelity*. In fact, *600* was just a week old when his sire, *Tutimes*, was basketed for the Rennes race in which he scored in the Scottish National.

Unmated Birds

In Chapter 6 I gave a list of the favourable nesting conditions in which the majority of pigeons produce their maximum endeavour. To every rule there is usually to be found an exception and some cocks, particularly yearlings, that have never been mated race very well in the unmated condition. When you find a pigeon of either sex performing well in the unmated condition, do not make the mistake of finding a mate for it hoping to improve its performance as the opposite is more likely to be the result. When things are going well there is nothing like sitting tight. I find the best way to treat unmated birds is to keep them in the YB or a separate compartment, exercise them with the mated birds and when fed in after exercise allow them a short spell among the mated pairs. Unmated hens will lay of course, and have to be carefully watched to ensure that they are not sent racing when about to lay. It is not advisable to send birds racing which have just deserted their eggs, for having lost interest they are likely to be stale. This applies equally to a bird which has just lost its mate. When the parents are racing never take the youngsters from them to wean until the parents are safely basketed. Many pigeons of both sexes race well when sitting a few days on eggs with a big youngster still in the nest and some fanciers think the attraction is the youngster. However, rarely does a big youngster without the eggs produce the desired result. The logic of the situation to me is that parents which lay quickly while still feeding youngsters, thereby give proof of their vitality and vigour, whereas those that are slow to lay agina suggest that the care of the youngsters has put a strain on their constitution.

Too Many Pigeons and Too Much Coddling

The signs are plentiful if we know where and what to look for and read the signs aright. Most of us keep far too many pigeons and an analysis of the season's racing will give ample proof of this. Just make a list of the birds which during the season's racing have been in the first three to the loft at least once; excluding young birds the rest could have been done without. Make another list consisting of pigeons which have bred for you even one bird which has survived two old bird racing seasons. When we get down to hard facts it is only the pigeons which appear on these lists that are really worth keeping. The late Jack Bruce was for thirty-six years secretary of the Manchester Flying Club and for fifty years the author of North-west Jottings in the *Racing Pigeon*, London. During a long lifetime he had endless opportunity to observe the methods of the Manchester cracks. His advice to young fanciers was, 'Don't coddle or nurse any racing pigeon, no matter what sort of pedigree or performance is bred in it, but give it the works. Only in very exceptional cases is any racing pigeon good enough to be kept for stock. I know of many instances of fanciers being ruined (not financially) by being persuaded to keep too many bought pigeons for stock. It is racing pigeons with guts most of us are after, pigeons that can keep on the wing for fourteen to sixteen hours the first day and carry on next morning and afternoon and be in time to win.' I know very well that there are those who will trot out instances of where a pigeon was not trained as a YB, got a 100 miles as a yearling, 250 miles or thereabouts as a two-year-old and in its third or fourth year won from 500 miles. My reaction to this is, what a waste of a good pigeon! In these instances we are not told of the dozens of pigeons treated this way and after years of coddling have let their owners down again and again. A pigeon's life is short and the sooner we find out its capabilities the better. Really good pigeons take all we ask of them in their stride; the best of them can make mistakes when their luck is out but give them the chance to recover and it rarely happens again. The reasons why we all keep too many pigeons are multifarious and I have no intention of attempting to list them here because there are only two good reasons for keeping any pigeon and that is to race or breed from. The racing part is simple, as Jack Bruce advises, give them the works. The worthwhile improve with this treatment and the others are simply not there and are soon forgotten. Selecting the birds to breed from is not so simple, but eventually the prepotent pigeons emerge, as shown by the survival of their progeny.

Reference Birds

Like our pigeons we all learn from experience and I well remember the first occasion I set about listing the pigeons of a deceased fancier for sale. I found in a book a list headed 'reference birds'. This list consisted of pigeons he had bought in or had been given to him by friends. There were perhaps a dozen of them and I set about typing the detailed pedigrees of these so-called 'reference birds' and after that I compiled the list of birds for sale. At the end of the day when I set about checking back from the birds for sale to the 'reference birds' only three out of the dozen were required. Breeding from the others had just been a waste of time and money. I was taught two lessons from this experience. The first, a purely personal one, was always make out the sale list first and the 'reference birds' will sort themselves out automatically. The second lesson drove home the fact that we persist year after year in breeding from pigeons that are just useless as producers. We keep birds for stock on the strength of their pedigree and the price we paid for them, together with many other futile reasons best known to ourselves. When it comes to breeding the only thing which matters is the kind of progeny a pigeon produces.

A Conversation Overheard

Before leaving this question of the number and kind of pigeons we keep, I would like to recount an experience I had many years ago. While sitting waiting for a meeting to commence I overheard the following conversation:

Fancier Green: 'I am looking for two hens to even up my pairs.'

Fancier Brown: 'Well, kill two cocks.'

Fancier Green: 'That is just the trouble, two of the hens I have are not much good.'

Fancier Brown: 'Well, kill those two hens and four cocks.'

I knew Fancier Brown only kept about eight pairs of pigeons but when it came to the races which mattered he was usually there or thereabouts with a good pooled pigeon. I had often wondered how he did it and I found the answer by overhearing the conversation which I have related.

The Backbone

It is often said that the ordinary fancier who pays his union fees is the backbone of the sport but to my way of thinking it is far from the truth. The backbone, nerves and sinews of the sport are the secretaries and the few other workers in every club who by their labours make it possible for others to enjoy their sport. They certainly do not do the work involved for the money they get, nor for honour and glory. There are a number known to me personally who have devoted their time and talents to the furtherance of the sport leaving behind a rich inheritance that is ours today. Talking over old times with my old friend Bob McKendrick, one of the early presidents of the Scottish National Flying Club, he recalled the day when the late James Wise of Uddington called on him. James Wise said, 'Come with me to Edinburgh, Bob. I am going to see a young chap called Dickson who might be suitable as secretary of the Scottish National Flying Club.' As the result of this the late Tom Dickson became secretary of the SNFC, a post he held for close on thirty years. Meticulous in all he did, a strict disciplinarian, a terror of the careless and forgetful, for him rules were rules and they had to be observed to the letter. In partnership with the late Dr. W. Anderson as President, from a small beginning the club grew in membership and prestige until it is the richest and best long-distance flying club in the world. The peak figures to date are over 2,600 members, sending over 6,000 pigeons to fly for some £27,000 in the Rennes race, and in the Nantes race £6,679. The club's most valued asset, however, is its prestige, and no one contri-tributed more to it than Tom Dickson. When he took over, the club had next to nothing in the way of funds. But he nursed it. Typical of the man was my seeing him arrive in Glasgow for the race marking and amongst other things handing to Jimmy Rankin the convoyer not a ball of twine to tie up the baskets, but the exact number of pieces required, cut to the necessary lengths. The late Jack Bruce I have already mentioned. His name was synonymous with the great Manchester Flying Club at its peak. A prodigious worker, he somehow contrived to see the funny side with a straight face which trapped the unwary. It was the late Fred Potts of Chester-le-Street whose undoubted organizing genius brought the Up North Combine to what it is today, with its 238 clubs of thirty Federations, requiring two special railway trains of approximately twelve pigeon vans in each to convoy an average of 20,000 each week during the racing season. In addition several aircraft are needed for their cross-Channel races. Then there is Alex Black, J.P., ex-Provost of his native Airdrie, whom I have known since we both joined the Airdrie and District Club in, I think, 1912. Of all the pigeon men I know, Sandy, as he preferred to be called, must be one of the most widely experienced of them all. Secretary of the Scottish Midland Federation since its inception forty years ago, he set an example over the years for quick results and payment of prizes which has come to be the standard expected of a good secretary. A tireless worker, he was an ex-President of the Scottish Homing Union and ex-Secretary and now President of the Scottish National Flying Club. An entertaining speaker, he was a popular guest at social functions during the winter. His down-to-earth approach to the everyday problems of a busy man is typified in the following incident. Approached by an irate fancier making a great noise over an alleged discrepancy in a race result, Sandy quietly said, 'Tell me all about it and if it is an error we will soon put it right.' Last but not least is my old friend Willie McClure of Prestwick, for thirty years secretary of the Kyle Federation and a councillor of the SHU. The weight of years had not dimmed his enthusiasm and wherever pigeon men meet, Willie would be there.

These are only a few of the many who have one thing in common—an intense interest in pigeon racing and all connected with it and workers among, I am afraid, many drones. Their counterparts are to be found wherever pigeon racing flourishes and I have this to say about each and every one of them: they are to be envied in that they find pleasure and happiness in doing so much for so many.

Sense of Values

Everyone is possessed of a given temperament which changes little with the passing years. The changes which take place are in our sense of values. This question of values is an important one because it is on his sense of values that the foundation of a fancier's success is based or on the other hand the rock on which he may come to grief.

Quite often when you are handed a pigeon for inspection you are told the price its owner paid for it or its parents, or are informed that it belongs to a popular strain of the day. This sort of thing may have impressed me at one time but not now. None of these things makes a pigeon better or worse as a breeder or racer; what counts is its own success as a breeder or racer and nothing else. Two of the most insidious and costly distorters of values are sentiment and favouritism; beware of these two deceivers. I have noticed, particularly with someone who has scored his first big success, that the winning pigeon becomes a little god and is raced no more. The old experienced hand puts a different value on a winning pigeon and races it. There is something which attracts one to a really good racer and you rarely forget them. Almost five years ago Alex Holgarth of the West of Scotland Club won the Dol race 500 miles of the Glasgow and District Federation when very few birds got through in race time. His winner was a less than medium size seven-year-old Blue cock, and I was assured that this pigeon in his eight seasons racing had not missed a race in the club's programme. This meant that it had taken part in some sixty races, including three from 500 miles. During this period he won several prizes including being twice first 250 miles and now at seven years old scored his greatest success. Another pigeon in the same category which comes to mind is Landles Brothers of Calderbank Red Hen, *Try Again*, SURP-51-M 1553. She flew 218 miles as a Y.B.; 365 miles as a Yearling in 1952; then Rennes 543 miles in 1953, 1954, 1955, 1956 and 1957, winning in 1957 6th Midland Fed. Open race 365 miles. In 1958, she won 74th West Sect., 129th Open Scottish National from Rennes and went on to fly Nantes 605 miles. In 1959, she won 33rd Sect., 88th Open Scottish Nantional Nantes. In 1960, at her ninth time across the Channel competing in the Scottish Nationals in their very hard race from Rennes, she came up to win 2nd Sect., 3rd Open, velocity 707.

Each year I visit where possible the winning lofts in Scottish National and other classic races, seeing, handling and taking factual details of the winning pigeons. The following table gives the nesting condition at basketing of these winners during the last two seasons.

TABLE 8

NESTING CONDITIONS

Class of winner	Hatching	SITTING (10 days) Under	Over	Small Youngster	Driving	Number
Outright winners	2	2	3	2	1	10
2nd Prize and section winners	1	5	4	—	1	11
Outstanding repeat winners	1	4	1	1	—	7
Totals	4	11	8	3	2	28

While the above figures are accurate they involve too few pigeons to provide a reasonable average. They are at least interesting. Note how the longest lasting conditions sitting provides 18 out of 28, but among the actual outright race winners only half of them were sitting.

I have also a record of the wing conditions of these twenty-eight pigeons over the period of the race, but as is to be expected these races were spread over four or five weeks with the moult going on all the time. This and whether a pigeon was a normal or slow moulter accounts for the fact that they varied from just casting their 2nd to growing their 5th. In not one single instance had a bird a full wing. Here again is something that will shake the wing faddist. One of the outright winners had cast its 2nd and 3rd flights in both wings a day or so before being basketed; among the Section winners one had cast its 3rd and 4th and another its 4th and 5th. Their ages ranged from a yearling to nine years old.

When a pigeon wins in a certain nesting condition, in preparing it for a repeat performance surely every effort should be made to send it in exactly the same condition as before; on the other hand, if it is a certain pigeon which looks the part but just keeps coming up from the longer races, it should not be discarded until it has been tried in other nesting conditions. Here are two examples of what I mean. In 1961 Leishman Brothers Red Hen 3730/59 *Holm Dewdrop* was sent to Nantes, with a day-old youngster to win 10th Section, 16th Open and £60; in 1962 she was sent to Rennes, sitting 11 days, timed but not placed. In 1963 and 1964 she was sent to Rennes, both times feeding a day-old youngster to win 48th/75th and £362; then 37th/74th and £373; a grand total of £795 in three Scottish National races.

Now take Duncan Ogilvie's Dark Chequer Pied Hen 1823/61 *Tibs*, one of the three hens which won for him 1st, 2nd and 3rd from Avranches. For three years she was tried sitting, but in 1965 she was sent with a week-old youngster and only then did she show her mettle from 500 miles.

Pigeons which will fly 500/600 miles in reasonable time are not so plentiful that we can afford to discard them, if at first they are not among the prize winners, nor on the other hand can we readily stop them on the strength of one success. Very few great racers are also great producers, so keep your racers racing, time enough to stop them when they have proved themselves at stock. The champion racers I have known became champions because they were kept racing, examples of which will have been found in the preceding pages of this book.

Do not be deluded; if your fancied pigeons do not stand the pace, they are not the right sort and/or there is something wrong with your management of them. As far as Scotland is concerned, and I do not suppose it differs greatly from other areas, so many put all their eggs in the one basket in a wild gamble in the big Rennes race, leaving themselves with neither the birds nor the money to compete in the races which follow. Shrewd, experienced fanciers play safe by preparing a number of birds to reach peak form between the end of June and mid-July, then selecting as each race comes along what they consider are the most likely of their candidates at that particular time. At a race marking, in conversation with an old friend, a shrewd and very successful fancier, in answer to my enquiry regarding what he was sending and how did he fancy his prospects, he replied: 'Andrew, sometimes we kid ourselves, but I don't think I'm kidding myself this time.' And he wasn't.

Pigeon racing is a strenuous sport demanding all we can put into it, yet take the average local club and after the 300-mile stage a surprising number of members have shot their bolt. It is true, some may send to the 350/400-mile stage but it is only on a rare occasion that one of them will be in the running. This is simply because they and their pigeons have not got what it takes. We have all experienced the dreadful feeling when, after a long afternoon and evening of patiently watching and waiting, darkness falls, then after a despairing look round we reluctantly lock up the loft with no birds home. This is the kind of medicine we must be prepared to swallow and then be up at the crack of dawn next morning ready for more. Three or four week-ends of this without success is a severe strain on our equanimity but it is all part of the sport of pigeon racing.

I wish the reader the success he deserves and I can assure him that when hope is fading with the daylight and then a lone pigeon drops to his loft everything else is forgotten in the thrill of that moment.

Appendix A

The 'Olympic' Method of Polygynous Matings

The need to increase the number of progeny which can be obtained from one individual pigeon in as short a time was possible, is a very real one, especially when seeking the answer to any problem concerning inheritance. Racing pigeon fanciers are naturally keen to obtain as many young as possible from an outstanding pigeon and resort to several methods to procure this end and six to eight youngsters in a season is considered good going. When it comes to some form of progeny testing by normal procedure several years may elapse before any concrete proof of a pigeon's capabilities as a producer are ascertained. In the average loft where a number of pigeons cannot be set aside purely for breeding or feeding purposes the more orthodox methods present many difficulties if the training and racing programmes of a number of pigeons are not to be upset. The 'Olympic Method' of producing a greater number of progeny from one cock has many advantages and the pigeons involved can be raced in the normal manner. It requires some skill in the management of the pigeons, plus an understanding of exactly what is happening at each stage.

Equipment

The only equipment required are nest boxes which are capable of being divided into two separate compartments, an inner and an outer one. The division between the compartments should be constructed of wooden dowels or wire rods, which enable the mated pairs to have a clear sight of each other.

It is not necessary that the pigeons concerned should be in a separate loft or compartment of the loft away from the other inmates, but if they can have a compartment to themselves so much the better.

Stud Cock

This is the pigeon from which you desire to produce the progeny, either because he is a proven producer of above average pigeons, or in order to test his capabilities as a producer.

Selected Hens

These are the hens from which you desire young sired by the stud cock.

Foster Cocks

These are the cocks which are to be used to stimulate the selected hens and later to sit and incubate the eggs. Having selected your stud cock and, for the purpose of this illustration, let us suppose you select five hens, the next step is to select four additional cocks (the foster cocks). This gives you five pairs of pigeons which you proceed to mate up in the normal way by placing them in their respective nest boxes, the hens in the inner compartment and the cocks in the outer. The cocks have their freedom to enter or leave the outer compartment but the hens are kept in the inner compartment in which you have provided water for them to drink. Watch each pair carefully to see if they are taking to each other. If you do have them in a separate compartment of the loft from the other birds, you can, next day, reverse the process by putting the cocks in the inner compartment and giving the hens their freedom in the outer compartment, thus getting them used to the position of their nest box. The main thing at this stage is to get each pair to take as soon as possible. If you have any doubts about any of the pairs leave them together taking care that no treading takes place. It is a good idea to let them have a run in the loft out of the nest box, again taking care that no treading takes place. This is particularly important if the other cocks are about as this encourages the cock to drive. During this period these birds are fed in their nest boxes.

Counting Days

This is an important point. The first day is twenty-four hours after the birds have taken to each other. You may put a pair of pigeons together and a day or so may elapse before they can truly be said to have taken to each other, and it is from the day you start counting. A careful note must be made of the exact date when each pair took to each other. The stud cock and his mate must be treated exactly the same as the other pairs. He can of course be allowed to tread his hen during these first three days. What you want to watch for is when he shows any signs of driving her to nest. He should normally be driving by the fourth counting day. On the evening of the third counting day, place the foster cocks in a basket and remove them from the loft. Now put the stud cock in the loft compartment and allow each of the selected hens in beside him and allow him to tread them, allowing a reasonable time to pass between each. The behaviour of the stud cock and each individual hen must be your guide to the success or otherwise of this manoeuvre. If a hen sits down quite readily without the usual courtship antics you have little to fear; from experience you will come to know when all is well. Do not expect too much on the third counting day. The fourth day is the critical time, as it is between the fourth and fifth day that the hen ovulates, at which time the descending ovum is fertilized and it is then that a hen will sit down to be trodden without the usual preliminaries. If you watch carefully, a hen on being trodden at this time gives a decided quiver of her wings at the moment of copulation. The fourth day then is the one on which you keep the stud cock busy. On the fifth day, as a safeguard, let the stud cock tread the selected hens again. Provided he is a vigorous cock in his prime and you have perhaps kept him away on his own during the third and fourth day, he will perform satisfactorily. In between your endeavours with the stud cock and the selected hens allow the foster cocks into their outer compartment with their hens safely in the inner compartments.

One or two of the selected hens may not have responded so well as the others and their counting days will therefore differ and they must be treated accordingly. On the sixth day place a nest bowl in each of the inner compartments, as the hens should lay the first egg on the seventh counting day. On the morning of the day after the first egg has been laid, mark it and put the egg in a safe place. Try to have the stud cock tread each hen on the day the first egg is due to be laid as this safeguards the fertility of the second egg. Late on the evening of the day before the second egg is due to be laid, take the partition out of the nest box thus allowing the foster cock free access to its mate. The foster cocks will be a bit boisterous at first, but by the morning they will have cooled down and will start driving the hen to nest and that evening she will lay her second egg. While she is sitting in the late evening replace the first egg. It will be unusual if each foster cock does not take to the eggs in the normal way. It is as well, however, to remove the hen from the loft or compartment while the foster cock is sitting, at least for the first day, just to see how he behaves. In due course, barring accidents, you should have ten youngsters from the one cock, with perhaps only a couple of days between the oldest and youngest.

The Reasons Why

Now for a word or two on the principles upon which this system is based. We know that the desire to mate is more or less dormant during the winter months. With the increasing hours of daylight in the spring the desire to mate is awakened. When you mate a pair of pigeons it is the excitement of the acts of courtship which brings the sex organs into a state of activity and a hen will lay without having been trodden. Thus by allowing these mated pairs to become excited over each other through the bars of the dividing partition, the mating sequence of events are set in motion. First, one of the tiny ova in the hen's ovary commences to develop rapidly until bursting its retaining membrane it drops into the oviduct, where it meets the sperm of the cock which punctures the ovum: thus the egg is fertilized. Experiments suggest that it is the latest and most active sperms which usually win the race to fertilize the descending ovum. In other words where a hen has been trodden by several cocks it is the sperm of the last cock to tread her prior to ovulation whose sperm will fertilize the egg. The success of the system is all a matter of timing and observation; all that matters is that the sperm of the stud cock is in the hen's oviduct when she ovulates. Some may worry about the possibility of some other cock treading the hen prior to her being mated or the foster cock doing so after mating. While it is much better that none of these things should happen, the evidence I have considered, on the basis of the inheritance of colour and plumage pattern, suggests that it is the last cock to tread the hen prior to ovulation that is the one most likely to sire the young. Thus the period during which the greatest care must be taken that no other cock treads a selected hen is from the third counting day until twenty-four hours before the laying of the second egg. This method must not be tried until the weather is good, say from mid-March onwards. Yearlings are usually uncertain and hard to time, so older birds are preferable as they will have been selected previously by racing. This system does not interfere with their racing and training as they go on just as if they had been mated in the usual way. A word of warning, however: do not try this with the same foster cocks in the next nest. Somehow they are not so easily manoeuvred into sitting the second time.

Alternative Method

The preparation and equipment are the same as the first method only the foster cocks are allowed to run with their hen naturally for the first three days; that is until they show signs of driving to nest. At this stage the selected hens are locked in the inner compartments and treated the same as in the first method from the fourth day onward.

General Use

The man with, say, twelve pairs can by this method have, say, four cocks sire the young of all his pairs. Each of these four stud cocks is given a mate and then manoeuvred to sire the eggs of three other hens. In fact, there is no end to its uses. I started this method many years ago in a modest way in order to get young from a good cock by two hens so that I could obtain a half brother and sister mating the following year. My method then was crude by comparison. It consisted of keeping a hen in a spare compartment in the loft and letting the stud cock in beside her every day until I found she was about to lay and then mating her to another cock I had kept spare for the purpose. Each evening I put this pair in the nest box together for a few minutes taking care there was no treading so that when I let them together I had no trouble. What I did not realize then was that while the first egg would be all right, the second egg might have been fertilized by the foster cock. Experience and increasing knowledge of the reasons why plus the urge for quicker results gradually brought about the improved technique as detailed above. As I have previously said, it requires a degree of skill in the management of the pairs concerned and a thorough knowledge of what you are about. If at first you do not succeed, try again. There are a few snags but none that cannot be overcome by patience.

Hints on Management

The foregoing instructions and principles of increasing the number of progeny from a stud cock cover this method where everything goes according to plan. There are a number of snags which may arise which have to be dealt with. The first of these is that a pair of pigeons may appear to have taken to each other through the bars of the dividing partition of the nest box, but may not behave that way when allowed together. It is for this reason that I suggest the pairs should be allowed together when you have time to watch them and see that no treading takes place. It also helps when you have two or three pairs mated to let them out together while you watch them; it encourages the cocks to drive their respective hens. The stud cock must also be encouraged to drive. It should be noted that a cock is less likely to drive if no other pigeons are around, so the more pigeons that are in the compartment or loft the better. Letting them out while you watch allows each pair to earn the way to their nest box and prevents fighting later on when the eggs are laid. The slow layer also gives trouble as the stud cock may be settled on his own eggs. Some knowledge of the inheritance of colour and pattern is a useful check as to which cock sired the eggs.

Advantages

In addition to producing a greater number of young from a selected stud cock, these youngsters are all of the same age and it is to be expected they will experience the same environment, training and racing. Thus we are in a better position to make comparisons between them in their performances. Furthermore, we have a number of youngsters from the same cock with different hens, giving us an early answer as to what hens the stud cock hits best. Where a young unproven cock is used this method can measure to a certain extent the young cock's capabilities as a producer. Either way it gives a quick answer which with normal methods would take years to obtain, not to mention that if the stud cock is prepotent we have a greater number of good pigeons in a very short space of time. With the young from a good sire there is a tendency to save them because in our estimation they are precious, but if we have a number of them we can afford to put them on the road racing and the survivors will, it is hoped, be the better pigeons. Those who practise inbreeding are by this method given a wider selection in their matings. In addition to all this, what appeals to me is that the pigeons used, even the stud cock, can go on with training and racing in the normal manner, and their nesting routine is not affected in the least. This method is better than mating a pair, letting the hen lay and then taking the eggs away to be reared by feeders. This practice upsets the moult of the birds involved and if they are intended for racing may upset the sequence of mating cycles in getting the birds ready for some of the longer events. Some may say it is asking a lot from the stud cock to tread all these hens, but just think again. I do not suppose that he will tread oftener by this method than if he were allowed to run free with his hen in the normal manner. With aged cocks it is a different matter and perhaps one or two hens in addition to his own mate will be enough; but it rests with our own judgement. Certainly try it with one or two hens to begin with and as you become more expert you can increase the number.

Appendix B

Sanitation in the Loft

The trite adage: 'Don't spare the scraper,' so often repeated by fanciers in their books and articles is to all intents and purposes a snare and a delusion. Racing pigeons are kept under artificial conditions and not in the natural environment of the species. The aim of the majority of fanciers is to maintain their pigeons as near to nature as possible; yet the first thing they do is to subject them to conditions of hygiene entirely opposite to their natural one. They scrape and scrub in an endeavour to maintain the standard of cleanliness of their own lofts only to find that moments afterwards droppings are all over the loft and their washing and scrubbing introduces the arch enemy of pigeons, dampness. The answer to this state of affairs is to introduce the *powdered droppings* method of loft sanitation. First of all please do not confuse '*Deep Litter*' as used by poultrymen, which consists of a mixture of various ingredients such as sawdust, peat and chopped straw, etc., and moisture which, in addition to the scraping among it by the fowl, has to be turned over regularly in conditions which raises the temperature of the house in which it is maintained.

Powdered droppings are exactly what their name implies and demand absolute dryness. Starting this method is most difficult; from my own experience and that of a few friends, the early difficulties are more psychological than real, for several reasons. The habits of perhaps a lifetime are not easily discarded and after years of scraping and cleaning it does not go down well at first to see your loft apparently looking untidy and neglected compared with the daily scraping days; and also to see the rising pile of droppings under the perches until there is an accumulation of dried droppings to absorb the moisture of fresh droppings. In consequence the new method may be abandoned before it has had a fair trial. Also, because of their structure, not all lofts are suitable to successfully practise this new method which demands good ventilation and perfect dryness. Steps must be taken to prevent rain and snow being blown in by the wind; overhanging trees will not help if they shade the loft from the beneficial rays of the sun shining on the loft.

Every loft will no doubt present its own problems, structural and otherwise; therefore in order to overcome the initial difficulties as already outlined a good start must be made. First collect the scrapings of droppings from your loft. They must not contain sawdust or other additives. Then spread them out and dry them, preferably in sunshine, as there is nothing like the intense drying effect of the sun's rays which also prevent the spread of many forms of lower life which are harmful to pigeons from spreading. When you have gathered enough of these powdered droppings to cover the floor and nest boxes of the loft to a depth of at least two inches, spread it out. These dried droppings readily absorb fresh droppings, thus accelerating their disintegration into powdered droppings. If, as is the common practice, you have your perches concentrated into groups, piles of droppings very quickly gather under the perches. My practice is to cover the droppings with powdered droppings where they may have gathered to excess in corners and spread them over these piles under the perches. I then break these droppings up and scatter them on the floor of the loft where they will quickly disintegrate in the powdered droppings.*

During the breeding season and moult, lightly rake the loft floor of straws used as nest-building material and of feathers and remove them from the loft. Do this about once a week. In recent years when visiting fanciers, particularly of the younger generation, who have been winning consistently, I have noted the number who have adopted this method of loft sanitation and I am sure that the majority will agree with me that a win-

* This method is not recommended if the fancier is susceptible to chest or breathing conditions.

ning loft must be a healthy one. I have also noted that after visiting one of these lofts I do not return home with tell-tale pigeon droppings sticking to my shoes.

Lastly, but of the utmost importance, protect your dried powdered droppings from spillings from the water fountains, using your ingenuity to achieve this, and do not allow your birds their bath inside the loft. Remember that complete dryness is the keynote.

Further information on loft sanitation the following publications will be of use:

1. CHAMBERLAIN, EDGAR, 1907. *The Homing Pigeon.* Manchester, England: The Homing Pigeon Publ. Co., p. 36.
2. LEVI, W. M., 1962 'Practical ideas about pigeons that are different.' *Amer. Pigeon Jour.* 15: 243 (June).
3. BUONFANTI, J. A., 1768. *Cel pollajo e della colombaja.* Livorno, Italy: Marco Cotellini in Via Grande.
4. FULTON, ROBERT and L. WRIGHT, no date 1875(?). *The Illustrated Book of Pigeons,* by Robert Fulton. London: Cassell Petter & Galpin, p. 371.
5. FULTON, ROBERT and W. F. LUMLEY, 1895. *Fulton's Book of Pigeons,* edited by Lewis Wright. London: Cassell & Co., Ltd., p. 506.
6. ARAGO, BUENAVENTURA (about 1900). *Tratado practico de la cria y multiplacion de las palomas.* Madrid: Hijos de Cuesta.
7. MACLEOD, A., 1913. *Pigeon Raising.* N.Y.: Outing Publishing Co., p. 68.
8. MACLEOD A., 1917. 'Don't clean too much.' *National Squab Magazine,* 8: 12, pp. 393–6.
9. LEVI, W. M., 1927. 'Questions and answers.' *Amer. Pigeon Jour,* 16:387 (Sept.).
10. LEVI, W. M., 1941. *The Pigeon.* 2nd ed. 1945, reprinted 1951, revised 1957.
11. SIEVERLING, J. W., 1942. 'Methods and management—sanitation.' *Amer. Pigeon Jour.* 31:200.
12. LEVI, W. M., 1946. *Making Pigeons Pay,* reprinted 1947, 1950, 1951, revised 1955 and subsequent reprint, p. 183 *et seq.*

13. KENNARD, D. C. and CHAMBERLIN, V. D., 1950. *Deep litter*. Ohio Agricultural Experiment Station, Wooster, Mimeograph Series No. 7.

14. LYELL, J. C., 1881. *Fancy Pigeons*. Second ed. 1887. Third ed. 1897. London: L. Upcott Gill., p. 23.

Appendix C

The Use of Cod Liver Oil

Cod liver oil, rich in vitamins A and D, should always be on hand in every pigeon loft as a body builder and tonic. In treating the individual pigeon recovering from an illness or from being flown down, the best means of administering its is by capsules. Empty capsules can be bought and filled as required with fresh cod liver oil as I have been told that cod liver oil loses its vitamin content when exposed to light and air. While I do not believe in pills and so-called tonics being given to otherwise healthy pigeons, I have found cod liver oil a useful and beneficial supplement to their normal diet. I find the best way of giving it is to take a tin which will hold more than enough food to provide a feed for all your birds, or one compartment if it is a big loft. Fill it up with beans or peas leaving enough room so that when you put on the lid the contents can be freely shaken. Now pour in a few drops of cod liver oil and shake the tin until all the beans or peas have a coating of cod liver oil. While the grains should all have a coating of cod liver oil, they should not have so much that they cling to one another; you will soon find out just how much oil is required. Now feed it to your pigeons. At first they may pause before eating the oil-coated grain but they will soon learn to take it without hesitation. I make it a practice to give my birds cod liver oil in this manner at the evening meal on Sundays and Wednesdays during the breeding and racing season. During the moult and when on winter rations, they get it three times per week with gratifying results. The tin in which you mix the grain and cod liver oil should be washed out occasionally with boiling water, otherwise the oil retained in the tin will become rancid.

Appendix D

The Problem of Strays

I suppose strays, like the poor, will always be with us and nowadays they present quite a problem for both the finder and the owner. Time was when the owner of a pigeon which had strayed could have it returned to him, if reported, for a matter of 15 pence of present-day currency. Today, however, I am doubtful, what with restricted railway services and other transport difficulties, if he could manage to have it returned to him for less than 100 pence, that is unless he could manage to collect it personally. However, what about the fancier into whose loft the wanderer has entered? The loft owner in these circumstances will be well advised to basket the wandering bird immediately and remove it, if at all possible, out of the sight and sound of his own birds. This is sound common sense for a number of very good reasons. First and foremost there is always the risk that it may be the vector in introducing one disease or another into your loft. It is true that racing pigeons are a hardy variety and as a general rule an ailing bird gets short shrift. The golden rule concerning strays is that on no account allow it to feed in your loft with your own pigeons. To do so will only encourage it to remain. If the bird concerned appears to be in fair physical condition, give it a light feed and water in a basket and in the morning, weather permitting, take it away and liberate it, taking care that your loft is shut up just in case it is inclined to return.

Never under-rate the facility pigeons possess of returning to locations which they have previously visited; the most obvious of these is their return when hungry to locations at which they have found food and/or a safe perch with the approach of darkness.

Standard Nomenclature

Phenotype	*Distinguishing Characteristics*
BARLESS BLUES BARLESS MEALIES (True)	These are the rarest of pattern pigeons and as their name implies they have no wing bars. The Blues have the black terminal tail bars found in all the Black/blue group and the Mealies the ash grey flights and tails of the Red/ash group.
BLUES and MEALIES (Barred)	There is no mistaking these birds with their two wing bars, Black and Red respectively. Note the width of the dove grey or ash grey band which separates the bars. This feature is helpful in differentiating between the pattern phenotypes.
LIGHT BLUE CHEQUERS	In both colours chequering will be found on the wings and on top of shoulders in the form of dark checks on a light background. The ragged lower edge of the bars reduces the area of the light band between the bars.
BLUE and RED CHEQUERS	With both these phenotypes the chequering gives the impression of light checks on a dark background. The wing bars are broad, thus reducing still further the light band be-the bars.
DARK CHEQUERS and DARK RED CHEQUERS	These are the darkest form of pattern, the light band between the bars has almost completely disappeared leaving only a thin

Phenotype	*Distinguishing Characteristics*
	line formed by tiny light ticks at the tip of the secondary cover feathers (coverts) with a few similar ticks scattered over the wing but there are none on the shoulder. They are described by some as Blacks, Velvets and 'T' pattern, or as Reds and Red Chequers.
BLACKS	These are easily recognized being completely black to the very down at the root of their feathers. An odd one may be found in which wing bars show through in a darker shade than the rest of its plumage.
ASH REDS	These are the Red/ash counterparts of Blacks and like the Blacks the odd one shows traces of bars on its wings. They are commonly known as Barless Mealies which they are definitely not.
RECESSIVE REDS	These are the only Self Reds and are recognized by the fact that their flights are red to their tips. Some will be found to have a few tail feathers of another colour.
DAPPLE—Pencil Blue, Strawberry Mealy	These are the birds known as Pencil Blues and their Red/ash counterparts Strawberry Mealies. Many chequers possess Dapple.
OPALS	Described by racing pigeon fanciers as Mosaics, their colour should be stated.
GRIZZLES	These are easily identified but a clearer picture is obtained by stating their colour and pattern.
PIEDS	These are pigeons with white about their head and neck and usually accompanied with several white flights.

Phenotype	*Distinguishing Characteristics*
BROWNS	These follow the same pattern markings as the Red/ash and Black/blue groups. They are often mistaken for Duns and Silver Chequers, and are difficult to recognize apart. Browns have light pearl eyes and the parts of their feathers exposed to sunlight fade.
DILUTES DUNS, Silver, Silver Chequers, Yellow, Yellow Chequers	Dilute is a sex-linked recessive characteristic which only some birds inherit. Its effect is to reduce the intensity of all three pigmentation (colour) groups, with the undernoted results. Blacks become Duns; Dark Chequers Dun Chequers; Barless Blues become Barless Silvers; Blues become Silvers (Barred). Light Blue Chequers become Light Silver Chequers; Chequers become Silver Chequers; these are all of the Black/Blue group. Dark Red Chequers become Dark Yellow Chequers; Light Red Chequers become Light Yellow Chequers; some of the light yellow birds are called Creams. When one of these birds unexpectedly turns up it is always a hen. These birds which inherit Dilute are recognizable at birth as they always have short down and look naked.

Glossary

ALLELOMORPHS. Two or more genes are said to be allelomorphs when they occupy the same relative position (loci) on homologous chromosomes. Several genes allelomorphic to each other are called alleles forming an allelomorphic series; not more than two members of a series of multiple alleles can simultaneously be present in the allelomorph of a normal diploid cell. They arise from mutations of one to another producing different forms of the same character.

AUTOSOME. A chromosome other than the sex chromosome. Autosomal Character: a character the genes determining which are carried on an autosome.

BACK CROSS. A mating between a homozygote and a heterozygote. Examples, sire to daughter, mother to son.

CELLS. The units into which the protoplasm is usually divided.

CHROMATIDS. One of the two longitudinally split chromosomes in the process of cell division in mitiosis or meiosis.

CHROMOSOME. One of the deeply staining paired bodies which appear in the nucleus preparatory to its division. They are generally constant in number in each species and carry the genes.

CROSSING-OVER. The mutal exchange of parts between homologous chromosomes during their pairing at meiosis.

CYTOPLASM. Protoplasm other than that contained in the nucleus of each cell.

DIPLOID. Individuals possessing two chromosomes of each kind in their body cells.

DOMINANT CHARACTER. One which is fully expressed whether the gene producing it its in the heterozygous or the homozygous state.

EPISTASIS. Where a character masks the effects of another character or characters the genes determining which are not alleles.

FACTOR. A Gene.

FERTILIZATON. The fusion of the male and female gametes, forming the zygote.

F1 GENERATION. The first filial generation, being the offspring of a given mating (parental generation).

F2 GENERATION. The second filial generation by interbreeding the F1 generation.

FIRST CROSS.The mating of two different species or between different breeds (varieties) of the same species. Also the mating of two closely inbred but unrelated individuals of the same variety and species. *See also* Heterosis.

GAMETE. The haploid reproductive cell of either sex (sperm of the cock, ovum of hens).

GENE. An hereditary factor. The genes are paired (allelomorphs) and are carried in paired (homologous) chromosomes, in which they are situated at definite loci.

GENETICS. The study of heredity and variation.

GERM CELLS. Cells specially set aside in the gonads for the production of the gametes.

GLANDS. Structures which manufacture substances required by the body.

GONADS. The organs which produce the gametes (testes of the cock and ovary of the hen).

HAPLOID. Cells containing only one of each kind of chromosome.

HEMIZYGOTE. An individual possessing only one chromosome of a pair. Hens are in effect hemizygous for the characters carried on their single Z chromosome as its homologous partner in the W chromosome is inactive. For instance Colour, Sex, Dilute and Short down.

HETEROGAMATIC SEX. Carrying unlike sex chromosomes (hens).

HETEROSIS. A characteristic of a first generation hybrid of producing one or more characters not expressed in either parent. Assumed to be due to special combinations of dominant genes usually applied to an increase in vigour, when it may be termed 'hybrid vigour'.

HETEROZYGOTE. An individual in which the members of a given pair of genes are dissimilar.

HETEROZYGOUS. When the genes in the allelomorph are dissimilar.

HOMOGAMATIC SEX. That carrying similar sex chromosomes (cocks).

HOMOLOGOUS CHROMOSOMES. The members of the same chromosome pair derived respectively from the two parents in bisexual reproduction.

HOMOZYGOTE. An individual in which the members of a given pair of genes are similar.

HOMOZYGOUS. When the genes in the allelomorph are similar.

HYBRID VIGOUR. *See* Heterosis.

LINKAGE. The tendency for genes to segregate together, instead of assorting independently because they are carried on the same chromosome pair (crossing-over can in certain circumstances offset the effects of linkage).

LOCUS (pl. Loci) The position occupied by a gene in (on) a chromosome, with regard to its linear order.

MEIOSIS. The process by which the nuclei of the cells set aside to produce the gametes, divide a second time with the result that the gametes (haploid) carry only one chromosome (chromatid) of each kind.

MITIOSIS. The process at cell division by which the nuclei of cells normally divide. It ensures that each daughter cell receives a longitudinal half (chromatid) of every chromosome.

MULTIPLE ALLELOMORPHS. A set of more than two genes, all allelomorphs (alleles) which have originated from each other by mutation.

MUTATION. A change in any unit of heredity. The majority are changes of individual genes, in which a gene while still influencing the same character as before does so in a different way. The changed gene is allelomorphic to the old.

NUCLEUS. A specialized part of the protoplasm of all normal cells, it carries the chromosomes.

OPTIMUM ENVIRONMENT. That best suited to the individual.

OVARY. The reproductive gland in which the ova are formed, it also produces female sex hormones.

OVIDUCT. The tube carrying the ovum from the ovary and in which the sperm of the cock meets and fuses with the descending ovum before it becomes encased with the white and egg shell.

OVULATION. The bursting of the ripe ovum from the ovarian follicle into the oviduct.

OVUM. The unfertilized egg cell (female gamete).

P1 GENERATION. The parental generation, consisting of the parents from which a breeding experiment starts.

PHENOTYPE. An individual judged from its appearance.

POLYPLOID. An individual possessing more than two sets of chromosomes.

POTENCE. The property of a group of polygenes corresponding to the degree of dominance of a major gene.

PREPOTENCY. The capacity of an individual to produce offspring more like itself than the other parent. Practical breeders' Term for a property depending (where genuine) on superior homozygosity, dominance and potence.

PROTOPLASM. The living substance of an organism.

PURE LINE. A group of individuals having all their allelomorphs homozygous.

RECESSIVE CHARACTER. One which is expressed only when the genes producing it are homozygous. An exception to this is in the hemizygous state in hens.

SEX CHROMOSOMES. The 'Z' ,'W' chromosomes in pigeons ('X', 'Y' in mammals).

SEX-LINKED GENES. Those carried on the 'X' chromosomes.

SIBS. Progeny of the same parents (full brothers and sisters).

SPERMATOZOON OR SPERM. The male or motile, reproductive cell (gamete).

TESTES. The organs which produce the male gametes (sperm): they also secrete male sex hormones.

WILD TYPE. Phenotype which is characteristic of the majority of individuals of a species in natural conditions.

'Z' CHROMOSOMES. The chromosomes carrying the genes which control the development of sex.

ZYGOTE. The fertilized ovum, the first diploid cell of the new individual before it undergoes further differentiation.

Index